Nonlinear Computer Modeling of Chemical and Biochemical Data

Nonlinear Computer Modeling of Chemical and Biochemical Data

James F. Rusling

Department of Chemistry
University of Connecticut
Storrs, Connecticut

Thomas F. Kumosinski

Eastern Regional Research Center
U.S. Department of Agriculture
Philadelphia, Pennsylvania

ACADEMIC PRESS
San Diego New York Boston London Sydney Tokyo Toronto

Copyright © 1996 by ACADEMIC PRESS, INC.

Academic Press, Inc.
A Division of Harcourt Brace & Company
525 B Street, Suite 1900, San Diego, California 92101-4495

United Kingdom Edition published by
Academic Press Limited
24-28 Oval Road, London NW1 7DX

Library of Congress Cataloging-in-Publication Data

Rusling, James F.
    Nonlinear computer modeling of chemical and biochemical data / by
James F. Rusling, Thomas F. Kumosinski.
            p.   cm.
    Includes bibliographical references and index.
    ISBN 0-12-604490-2 (alk. paper)
    1. Chemistry--Statistical methods--Data processing. 2. Regression
analysis--Data processing. 3. Nonlinear theories. I. Kumosinski,
Thomas F. II. Title.
QD39.3.S7R87   1996
540'.1'175--dc20   95-30376
                                              CIP

PRINTED IN THE UNITED STATES OF AMERICA
96   97   98   99   00   01   BB   9   8   7   6   5   4   3   2   1

We dedicate this book to the memory of Professor Louis Meites, who was ahead of his time as an innovator in the analysis of chemical data with computers

Contents

PART II
Selected Applications

Preface

On a warm Connecticut evening in 1989, the two coauthors of this volume discussed the lack of books that explain computer modeling methods, in particular nonlinear regression analysis, to chemists and biochemists in an understandable way. Our own early experiences in this field required digging out information from very good but mathematically dense texts. Thus, we felt that there was a need for a monograph introducing the practical aspects of nonlinear regression analysis to the average chemist and biochemist.

This book is the result of our discussions. We attempt to provide a solid introduction to the basics of nonlinear regression analysis in a format that does not rely heavily on mathematical symbolism. Only a rudimentary knowledge of algebra, elementary calculus, statistics, and computers should be necessary for the reader. The book is suitable for senior undergraduates and graduate students in chemistry or biochemistry, as well as professional practitioners in these fields. Part of the material has been taught to senior chemistry majors and graduate students in courses on computers in analytical chemistry for 15 years.

For those with more advanced mathematical skills, ample references to source statistical literature are given in Chapters 1 to 4. These chapters provide a general introduction to the subject. Near the end of Chapter 2, we list current sources of software for nonlinear regression.

We present detailed models, examples, and discussions of applications to specific experimental techniques in Chapters 5 to 14. They include potentiometric titrations, macromolecular equilibria, X-ray scattering, ultracentrifugation, electroanalytical methods, and spectroscopic and chromatographic analyses. In most cases, it should be relatively easy for readers to adapt the models in these chapters to their specific applications. We hope that this book fulfills our goal of enabling new or tentative users to model their experiments as painlessly as possible.

James F. Rusling
Thomas F. Kumosinski

Acknowledgments

We acknowledge the assistance, understanding, and forbearance of our families and our colleagues during the writing of this book, which necessarily took us away from many other projects. Special thanks are extended to Penelope Williams for her patience and kindness with both of us and to Joe Unruh (U.S. Department of Agriculture), who tirelessly and expertly prepared many of the figures for this book. We also are greatly indebted to all of our colleagues and students who contributed to the development of many of the concepts discussed on the following pages.

We thank the graduate students, staff, and postdoctoral fellows in the Electrochemistry Group of the Chemistry Department, Loughborough University of Technology, Loughborough, UK, where the majority of this book was written while one of us was on sabbatical. Specifically, we thank Dr. Roger Mortimer, Dr. Phil Mitchell, and Professor Frank Wilkenson of the Chemistry Department at Loughborough for their kindness and companionship and for making the full facilities of their department available to us. We also thank Simon Foster and Joe Wallace, who were always willing to help out with technical difficulties and cheer up us yanks with humorous comments.

For their interest and contributions of unpublished data, we thank Dr. James D. Stuart, University of Connecticut, and Drs. Harry Farrell and Mike Alaimo, of the U.S. Department of Agriculture. We are grateful to Drs. De-Ling Zhou and Qingdong Huang, who contributed both data and regression analysis results for Chapter 11, and to Dr. Artur Sucheta, University of California at Santa Cruz, who was kind enough to write a section of Chapter 4 on error surfaces and to proofread much of the text.

James F. Rusling
Thomas F. Kumosinki

PART I

General Introduction to Regression Analysis

Chapter 1

Introduction to Nonlinear Modeling of Data

A. What Is Nonlinear Modeling?

Fitting models to data is central to the quantitative pursuit of modern science. Let us first consider what types of situations involving measurements might arise during a typical day's work in a laboratory. Suppose I want to measure the pH of an amino acid solution. I place a precalibrated pH electrode into the solution and shortly thereafter read its pH on the pH meter to which the electrode is connected. This type of simple "read out" experiment is an exception. Few instrumental methods in chemistry and biochemistry can directly give us the specific information we set out to find.

In the most modern instrumental methods, the raw data obtained must be analyzed further to extract the information we need. Fortunately, this task can be handled automatically by a computer. For example, suppose we want to measure the diffusion coefficient of a protein in aqueous solution to get an idea of its molecular size. We have no "diffusion meter" available for this job. We need to do an experiment such as ultracentrifugal sedimentation, dynamic light scattering, or pulsed-field gradient spin-echo nuclear magnetic resonance and extract the diffusion coefficient by analyzing the data appropriately. Similarly, if we wish to know the lifetime of a transient species in a chemical or photochemical process, we cannot measure this quantity directly. The required information can be obtained by forming the transient species and measuring its decay with time. The resulting data are analyzed mathematically to obtain the lifetime.

Thus, extraction of meaningful parameters by computer analysis of experimental data is a widely required procedure in the physical sciences. The aim of this book is to describe general methods to analyze data and extract relevant information reliably in situations similar to those just described. These methods involve analyzing the data by using appropriate models. The models should be accurate mathematical descriptions of the response measured in the experiment that include all relevant contributions to the resulting signal. The goal of the computer data analysis is to reveal the relevant information in the raw experimental data in a form that is useful to the experimenter.

Computer modeling of experimental data, as the phrase is used in this book, involves using a computer program that can fine tune the details of the model so that it agrees with the data to the best of its ability to describe the experiment. Procedures in this computer program are designed to provide the best values, in a statistical sense, of numerical parameters in the model. These parameters might include diffusion coefficients or lifetimes discussed in the previous examples. In some cases, parameters will need to be interpreted further to obtain the desired information. For example, we might estimate decay lifetimes for a wide range of experimental conditions to elucidate a decay mechanism. A general block diagram of the analysis procedure is shown in Figure 1.1.

The computer programs utilized for modeling data should provide statistical and graphical measures of how well the model fits the data. These so-called goodness of fit criteria can be used to distinguish between several possible models for a given set of data [1, 2]. This might be the specific goal of certain analyses. In fact, goodness of fit parameters and graphic representations of deviations from models can be used as the basis for an expert system; that is, a computer program that finds the best model for sets of data on its own [1].

An example of a simple linear model familiar to chemists is Beer's law, which describes the absorbance (A) of light by a solution containing an absorbing molecule. The measured absorbance is linearly related to the product of the concentration of the absorber (C), the path length of light through the sample (b), and the molar absorptivity (ε) of the absorber:

$$A = \varepsilon bC. \qquad (1.1)$$

This model is linear, and linear regression analysis [3] can be used to obtain the molar absorptivity for a system with a known path length. The data

Figure 1.1 A typical analysis procedure.

analyzed are the measured absorbance vs. the concentration of absorber. For such linear models, closed form equations enable rapid, one-step calculations of the parameters of the model, such as the slope and intercept of a straight line. In this example, the slope of the regression line is directly proportional to the molar absorptivity.

Unfortunately for us, nature tends to be nonlinear. The preceding scenario involving a linear model is only a special case. Many models for experimental data turn out to be nonlinear. An example is the transient lifetime experiment discussed previously. If the decay of the transient species is first order, the rate of decay is proportional to the amount of transient, and the signal decreases exponentially with time. Also, at a given time, the signal decreases exponentially as the lifetime decreases. The model is nonlinear. Linear regression cannot be used directly to obtain lifetimes from the intensity vs. time data.

Although, linearization of a nonlinear model can be an option for analyzing data, it is often unwise. Reasons for this will be discussed in Chapter 2, where we also show that nonlinear models can be fit quite easily to data in a general way by using *nonlinear regression analysis.* The same principles apply as in linear regression (linear least squares), and in fact the two methods have a common goal. However, in nonlinear regression, we shall see that a stepwise, iterative approach to the least squares condition is necessary [4, 5].

B. Objectives of This Book

In the following chapters, we discuss the fundamental basis, procedures, and examples of computer modeling of data by nonlinear regression analysis. We hope to do this in a way that any interested practitioner of chemistry or biochemistry will be able to understand. No advanced mathematical training should be necessary, except for understanding the basics of algebra and elementary calculus.

Nearly all of the methods and applications in this book are suitable for modern (circa 1995) personal computers such as IBM-PCs, their clones, and Macintosh machines. A beginning-to-intermediate level of microcomputer literacy is necessary to use the methods described. An intermediate-level familiarity with computer programming in BASIC or FORTRAN or some knowledge of a suitable general mathematics software package (see Chapter 2) is required for application of the techniques discussed. Although a variety of programming languages could be used, we have focused on BASIC and FORTRAN because of the traditional use and familiarity of these languages in the scientific community. We have also presented a few applications in the Mathcad environment (see Chapter 2) because it is easy to learn and use and is familiar to us.

We now give a brief preview of what is to follow. Chapters 2 to 4 discuss general aspects of using nonlinear regression. Chapter 2 discusses the nature of linear and nonlinear models and algorithms and the operations involved in linear and nonlinear regression. It also provides a brief source list of appropriate mathematics software packages. Chapter 3 is a detailed tutorial on how to construct suitable models for analysis of experimental data. Chapter 4 discusses approaches to solving the problem of correlation between parameters in models and other difficulties in nonlinear regression analysis. This chapter makes use of the concept of graphical error surfaces and shows how their shapes can influence the quality of convergence.

Chapters 5 to 14 present specific selected applications of computer modeling to various experiments used in chemical and biochemical research. Highlights of these chapters include the determination of analyte concentrations by titrations without standardizing the titrant, estimation of electron transfer rate constants and surface concentrations from electrochemical experiments, and the estimation of the secondary structure of proteins from infrared spectroscopic data. These applications chapters include a short review of principles and models for each technique, examples of computer modeling for real and theoretical data sets, and selected examples from the literature specific to each particular instrumental technique.

The examples in Chapters 5 to 14 have been chosen for their tutorial value and because of our own familiarity with the instrumental methods involved. There are many more excellent research applications involving computer modeling of data, which we have not had room to include in this book. We have tried to limit ourselves to illustrative examples compatible with modern personal computers (i.e., circa 1995). We realize, however, that the microcomputer revolution is ongoing. The future should bring to our desktops the capability of analyzing increasingly larger data sets with increasingly more complex models. The reader is directed to several review articles for a taste of future possibilities [6, 7].

References

1. L. Meites, "Some New Techniques for the Analysis and Interpretation of Chemical Data," *CRC Critical Reviews in Analytical Chemistry* **8** (1979), pp. 1–53.
2. J. F. Rusling, "Analysis of Chemical Data by Computer Modeling," *CRC Critical Reviews in Analytical Chemistry* **21** (1989), pp. 49–81.
3. P. R. Bevington, *Data Reduction and Error Analysis for the Physical Sciences.* New York: McGraw-Hill, 1969.
4. Y. Bard, *Nonlinear Parameter Estimation.* New York: Academic Press, 1974.
5. D. M. Bates and D. G. Watts, *Nonlinear Regression Analysis and Its Applications.* New York: Wiley, 1988.
6. R. L. Martino, C. A. Johnson, E. B. Suh, B. L. Trus, and T. K. Yap, "Parallel Computing in Biomedical Research," *Science* **265** (1994), pp. 902–908.
7. L. Greengard, "Fast Algorithms for Classical Physics," *Science* **265** (1994), pp. 909–914.

Chapter 2

Analyzing Data with Regression Analysis

A. Linear Models

As an introduction to nonlinear regression, we start with a review of linear regression analysis. The main characteristic of linear models is that the measured quantity is linearly dependent upon the *parameters* in the model. Beer's law, mentioned in Chapter 1, is a linear model. If we measure the absorbance A at various concentrations C_j of an absorbing chemical species, this model has the form

$$A = b_1 C_j. \tag{2.1}$$

The only parameter in this model is b_1, the product of the cell thickness and the molar absorptivity (cf. eq. (1.1)). We shall see later that it is quite important in computer modeling of data to account for background signals. In this case, let us assume that we have a constant background signal b_2 derived from the solution in which the absorber is dissolved. The model for the Beer's law experiment becomes

$$A_j = b_1 C_j + b_2 \tag{2.2}$$

which is an equation for a straight line. In the interest of generality, we will convert A in eq. (2.2) to y, called the *measured* or *dependent variable*, and convert C to x, which is the experimentally controlled or *independent variable*. Including the constant background, eq. (2.1) can be written in the well-known form of a straight line:

$$y_j = b_1 x_j + b_2. \tag{2.3}$$

Equation (2.3) is a linear model because the independent variable y is a linear function of the model parameters b_1 and b_2, not because the

Table 2.1 Examples of Linear Models

Linear model	Equation number
$y_j = b_1 x_j^3 + b_2 x_j^2 + b_3 x_j + b_4$	(2.4)
$y_j = b_1 \log x_j + b_2$	(2.5)
$y_j = b_1 \exp(x_j^2) + b_2 x_j + b_3$	(2.6)
$y_j = b_1/x_j^{1/2} + b_2$	(2.7)

equation describes a straight line. The model need not be a straight line to be linear. Some other linear models are listed in Table 2.1.

Note that none of the linear models in eqs. (2.4) to (2.7) describe a straight line. The first is a polynomial in x, the second depends logarithmically on x, the third depends exponentially on x^2, and the last one depends on $1/x^{1/2}$. However, in all of these equations, the dependent variable y_j depends linearly on the k parameters $b_1, \ldots, b_k$. These parameters appear raised to the first power in all the eqs. (2.3) to (2.7).

Thus, we can have models that are not straight lines but are still considered linear for regression analyses. The preceding models have only one independent variable, x. They are called *single equation* models. Nowhere in these linear models do we see a term such as b_1^2, $\exp(b_1 x)$, or $\log(b_1 x)$. This would make the models nonlinear, because y_j would depend on one of the parameters in a nonlinear fashion. We can also have linear models containing more than one independent variable.

A.1. Linear Regression Analysis

The classification of a model as linear means that we can fit it to experimental data by the method of linear regression. Although we shall be concerned mainly with nonlinear models and nonlinear regression in this book, it is instructive to review the method of linear regression, also called *linear least squares.* We shall see that the same principles used in linear least squares also apply to nonlinear regression analysis.

Linear least squares is familiar to most scientists and students of science. To discuss its principles, we express a single-equation linear model with k parameters in a general form:

$$y_j(\text{calc}) = F(x_j, b_1, b_2, \ldots, b_k) \qquad (2.8)$$

where the $y_j(\text{calc})$ depend linearly on the parameters $b_1, \ldots, b_k$. Equation (2.8) provides a way to compute the response $y_j(\text{calc})$ from the linear model. We will also have n experimentally measured values of $y_j(\text{meas})$, one at each x_j. For a set of n measured values $y_j(\text{meas})$, we define the *error sum,* S, as

$$S = \sum_{j=1}^{n} w_j [y_j(\text{meas}) - y_j(\text{calc})]^2 \tag{2.9}$$

where the w_j are weighting factors that depend on the distribution of random errors in x and y. The simplest and most often used set of assumptions is as follows: the x values are free of error, and the random errors in y are independent of the magnitude of the $y_j(\text{meas})$. Another way of expressing this error distribution is that the variances in all of the y_j (σ_y^2, where σ_y is the standard deviation in y) are equal. In this special case $w_j = 1$. Other important options for w_j will be discussed later.

The principle of least squares is used in both linear and nonlinear regression. Its major premise is that the best values of the parameters $b_1, \ldots, b_k$ will be obtained when S (eq. (2.9)) is at a minimum value with respect to these parameters. Exact formulas for the parameters can be derived for a linear model by taking the first derivative of S with respect to each parameter and setting each of these derivatives equal to zero. This results in a set of linear simultaneous equations that can be solved in closed form for the unique value of each parameter [1, 2].

Consider this procedure applied to the straight line model in eq. (2.3):

$$y_j = b_1 x_j + b_2. \tag{2.3}$$

With the assumption $w_j = 1$, the form of the error sum is

$$S = \sum_{j=1}^{n} [y_j(\text{meas}) - b_1 x_j - b_2]^2. \tag{2.10}$$

For a given set of data, S depends on parameters b_1 and b_2. The required derivatives are

$$\partial S/\partial b_1 = 0$$
$$\partial S/\partial b_2 = 0. \tag{2.11}$$

Equation (2.11) yields two equations in two unknowns, which can be solved for b_1 and b_2. If we define x_{av} and y_{av} as

$$x_{av} = \sum_j x_j/n \tag{2.12}$$
$$y_{av} = \sum_j y_j/n. \tag{2.13}$$

The resulting expression [2] for the slope of the line is

$$b_1 = \frac{\sum_j [(x_j - x_{av})(y_j - y_{av})]}{\sum_j [x_j - x_{av})^2} \tag{2.14}$$

and the intercept is

$$b_2 = y_{av} - b_1 x_{av}. \tag{2.15}$$

The standard deviations in the slope and intercept can be computed from simple formulas [2]. A BASIC program for linear regression using a straight line model is listed in Appendix I. Also, built-in functions in mathematics software such as Mathcad, Mathematica, and Matlab can be used for linear regression. The book by Bevington [3] is an excellent source of linear regression programs (in FORTRAN) for a variety of linear models.

An important question to be answered as a part of any regression analysis is, How good does the model fit the data, or what is the *goodness of fit*? The product–moment correlation coefficient, often called simply the *correlation coefficient,* is a statistic often used to test goodness of fit of linear least squares models to data. This correlation coefficient (r) is defined as

$$r = \frac{\sum_j [(x_j - x_{av})(y_j - y_{av})]}{\{[\sum_j (x_j - x_{av})^2][\sum_j (y_j - y_{av})^2]\}^{1/2}} \tag{2.16}$$

and has values from -1 to $+1$. A value of $r = +1$ demonstrates a perfect correlation between x and y. Perfect negative correlation is denoted by $r = -1$. For calibration plots in analytical chemistry, using Beer's law data, for example, we generally would like to have $r > 0.99$. For this type of calibration data, values of $\mathbf{r} < 0.90$ would usually suggest an unacceptably poor correlation between y and x.

We now examine a typical linear regression analysis of Beer's law data. The model used is eq. (2.2):

$$A_j = b_1 C_j + b_2 \tag{2.2}$$

with $b_1 = \varepsilon l$, where path length l is 1 cm, and ε is the molar absorptivity in L mol^{-1}cm^{-1}. The data represent a series of absorbance measurements A_j over a range of micromolar concentrations C_j of an absorber. The results of analysis of these data by the BASIC program for linear least squares in Appendix I are given in Table 2.2 This program makes the standard assumptions that random errors in y are independent of the value of y(meas) and that there are no errors in x. This corresponds to an unweighted regression, with $w_j = 1$ in the error sum S (eq. (2.10)).

Least squares analysis of the experimental data in the first two columns of Table 2.2 gave a slope of $9.94 \pm 0.05 \times 10^{-3}$. Because the path length is 1 cm and the concentration units are μM, we multiply by 10^6 to convert the slope to $\varepsilon = 9{,}940$ L mol^{-1}cm^{-1} for the absorber. The intercept is $1.24 \pm 1.51 \times 10^{-3}$. The standard deviation of the intercept is larger than the intercept itself, suggesting that it is not significantly different from zero. A Student's *t-test* comparing the intercept to zero can be done to confirm this conclusion statistically [2].

Note that the correlation coefficient in Table 2.2 is very close to one. This confirms an excellent correlation of y with x. Because of the assumptions we made about the distribution of errors in y and the lack of errors in x, the

Table 2.2 Linear Regression Analysis of Beer's Law Data

x_j (μM)	y_j(meas) (A)	y_j(calc) (A)	Difference y_j(meas) $-$ y_j(calc)	Diff./SDR[a]
1.0	0.011	0.01118	-1.84E-04	-0.0754
5.0	0.049	0.05096	-1.96E-03	-0.803
10.0	0.102	0.10068	1.32E-03	0.541
20.0	0.199	0.20013	-1.13E-03	-0.463
30.0	0.304	0.29957	4.43E-03	1.81
40.0	0.398	0.39901	-1.01E-03	-0.414
50.0	0.497	0.49846	-1.45E-03	-0.595

Slope = 9.944367E-03
Standard deviation of the slope = 5.379905E-05
Intercept = 1.239661E-03
Standard deviation of the intercept = 1.51158E-03
Correlation coefficient (**r**) = 0.9999269
[a] Standard deviation of the regression = 2.438353E-03

standard deviation of the regression (SDR) is expressed in the units of y. To interpret this statistic in terms of goodness of fit, we compare it to the standard error (e_y) in measuring y. In this case, e_y is the standard error in A. Suppose we use a spectrophotometer that has an error in the absorbance $e_A - \pm0.003$, which is independent of the value of A. (This error would not be very good for a modern spectrophotometer!) We have SDR = 0.0024 from Table 2.2. Therefore, SDR $< e_A$, and we consider this as support for a good fit of the model to the data.

Although summary statistics such as the correlation coefficient and SDR are useful indicators of how well a model fits a particular set of data, they suffer from the limitation of being *summaries*. We need, in addition, to have methods that test each data point for adherence to the model. We recommend as a general practice the construction of graphs for further evaluation of goodness of fit. The first graph to examine is a plot of the experimental data along with the calculated line (Figure 2.1(a)). We see that good agreement of the data in Table 2.2 with the straight line model is confirmed by this graph. A very close inspection of this type of plot is often needed to detect systematic deviations of data points from the model.

A second highly recommended graph is called a *deviation plot* or *residual plot*. This graph provides a sensitive test of the agreement of individual points with the model. The residual plot has [y_j(meas) $-$ y_j(calc)]/SDR on the vertical axis plotted against the independent variable. Alternatively, if data are equally spaced on the x axis, the data point number can be plotted on the horizontal axis [1]. The quantities [y_j(meas) $-$ y_j(calc)]/SDR SDR are called the *residuals* or *deviations*. They are sometimes given the symbols dev_j. The dev_j are simply the differences of each experimental

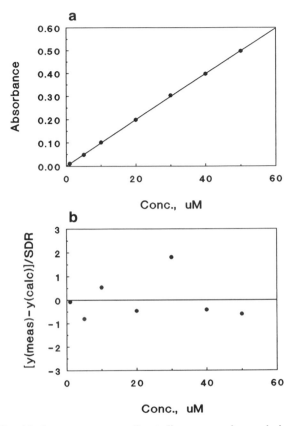

Figure 2.1 Graphical output corresponding to linear regression analysis results in Table 2.2: (a) points are experimental data, line computed from regression analysis; (b) random deviation plot.

data point from the calculated regression line normalized by dividing by the standard deviation of the regression. Division by SDR allows us to plot all deviation plots on similar y-axis scales, which are now scaled to the standard deviation in y. This facilitates comparisons of different sets of experimental data.

In the example from Table 2.2, the residual plot (Figure 2.1(b)) has points randomly scattered around the horizontal line representing $dev_j = 0$. This random scatter indicates that the model provides a good fit to the data within the confines of the signal to noise ratio of the measurements. This type of plot is a highly sensitive indicator of goodness of fit.

So far, we have discussed only data that agree very well with the chosen model. We now explore a second set of data that are fit less well by the linear Beer's law model. The data in Table 2.3 gave a slope of

Table 2.3 Linear Regression Analysis of Data with a Slight Curvature onto a Straight Line Model

x_j (μM)	y_j(meas) (A)	y_j(calc) (A)	Difference y_j(meas) $-$ y_j(calc)	Diff./SDR[a]
1.0	0.021	0.02953	-8.54E-03	-1.129
5.0	0.067	0.06928	-2.28E-03	-0.300
10.0	0.126	0.11895	7.05E-03	0.929
20.0	0.220	0.21829	1.71E-03	0.225
30.0	0.325	0.31764	7.36E-03	0.970
40.0	0.421	0.41698	4.02E-03	0.530
50.0	0.507	0.51633	-9.32E-03	-1.227

Slope = 9.934478E-03
Standard deviation of the slope = 1.676636E-04
Intercept = 1.960314E-02
Standard deviation of the intercept = 4.710805E-03
Correlation coefficient (**r**) = 0.9992887
[a] Standard deviation of the regression = 7.590183E-03

$9.93 \pm 0.16 \times 10^{-3}$ when fit to the straight line model in eq. (2.2). As before, this gives ε = 9,930 L mol^{-1}cm^{-1} for the absorber. The intercept is $1.96 + 0.47 \times 10^{-2}$, which in this case suggests that there is a real positive intercept. The correlation coefficient is 0.9993, which might lead us to believe that the fit is good. However, one indication that there may be a problem with the fit of the model to these data is that SDR is 0.008, which is greater than our estimate of 0.003 for the error in A. Another subtle indicator of a poor fit is that the deviations in the last column of Table 2.3 seem to follow a trend; that is, two negative residuals are followed by four positive ones then a final negative value.

Plots of the data in Table 2.3 can be used to support more solid conclusions about a rather poor fit to the model. Careful inspection of the calibration graph (Figure 2.2(a)) reveals that the first and last data points are slightly below the calibration line, while points 3–6 are slightly above the calibration line. The residual plot (Figure 2.2(b)) provides the same information in a clearer format. We see that the dev$_j$ are arranged roughly in the pattern of an upside down parabola. This evident pattern in the residual plot, or *deviation pattern,* indicates that the model does not fit the experimental data well. There can be a number of reasons for this, and they will be discussed later.

Table 2.3 and Figure 2.2(b) illustrate a case where summary statistics provide somewhat confusing evidence concerning goodness of fit. The value of **r** suggests a good correlation between y and x, but RSD is somewhat larger than the errors in y. However, small systematic deviations of the data from the model are clearly apparent from the residual plot (Figure

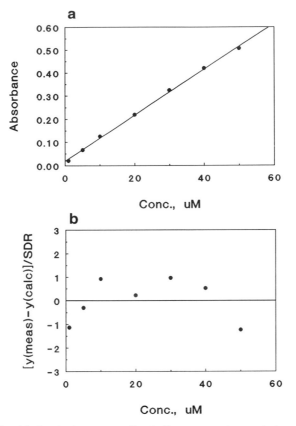

Figure 2.2 Graphical output corresponding to linear regression analysis results in Table 2.3: (a) points are experimental data, line computed from regression analysis; (b) nonrandom deviation plot.

2.2(b)). From consideration of all these criteria, we conclude that the model fits the data poorly. The residual plot is a major tool is establishing the poor quality of the fit. We will return to this theme often in later discussions of nonlinear regression analysis.

Residual plots can also reveal trends in the errors in y_j. For example, the plot in Figure 2.3 shows nearly random scatter about $dev_j = 0$, but the envelope of positive and negative deviations increases with increasing x. This type of deviation plot suggests that the original assumption about the constant variance in y is incorrect [3]. The plot suggests that the errors in y depend on the magnitude of y, because the size of the deviations of y from the model increase with x. As discussed in Sections A.3 and B.5, the $w_j = 0$ assumption cannot be used in such cases, and the regression analysis needs to be weighted appropriately.

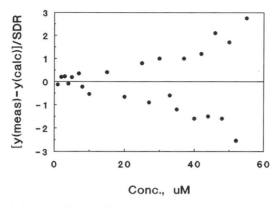

Figure 2.3 Deviation plot characteristic for an unweighted regression analysis when the errors in y depend on the size of y.

A.2. Matrix Representation of Linear Least Squares

It is usually more convenient to base programs for nonlinear regression on matrix algebra. This is the approach taken in higher level mathematical programming software such as that provided by Matlab and Mathcad (sources for the Matlab and Mathcad software are listed at the end of this chapter). The principles are exactly the same as in the algebraic approach discussed above, but matrix methods facilitate organization and manipulation of the data.

In matrix notation, the straight-line model can be expressed as [3, 5]

$$\mathbf{Y} = \mathbf{Xb} + \mathbf{e} \tag{2.17}$$

where $\mathbf{Y}$ is a vector containing the n values of $y_i(\text{meas})$, $\mathbf{X}$ is an $n \times 2$ sample matrix, $\mathbf{e}$ is a vector containing the observed residuals, and $\mathbf{b}$ is the vector containing values of the slope and intercept. For an example with $n = 3$, eq. (2.17) can be represented as in Box 2.1.

$$\mathbf{Y} = \mathbf{Xb} + \mathbf{e}$$

$$\begin{bmatrix} y_1 \\ y_2 \\ y_3 \end{bmatrix} = \begin{bmatrix} 1 & x_1 \\ 1 & x_2 \\ 1 & x_3 \end{bmatrix} \times \begin{bmatrix} b_1 \\ b_2 \end{bmatrix} + \begin{bmatrix} e_1 \\ e_2 \\ e_2 \end{bmatrix}$$

Box 2.1 Elements of vectors and matrices in a straight-line model.

Using the same assumptions as in the algebraic approach to obtaining the parameters, we have variances in y_j(meas) that are all equal, and no errors in x_j. In the matrix notation, the problem involves finding the parameter matrix **b**, whose elements are the slope b_1 and the intercept b_2. Application of the least squares condition leads to estimation of **b** from [4, 5]

$$\mathbf{b} = [\mathbf{X'X}]^{-1}\mathbf{X'Y} \tag{2.18}$$

where $\mathbf{X'}$ is the transpose of $\mathbf{X}$, and $[\mathbf{X'X}]^{-1}$ is the inverse of the matrix product $\mathbf{X'X}$. Efficient methods of computing **b** are discussed by Bates and Watts [5]. Also, Bevington's book [3] provides a good discussion of the matrix approach starting from an algebraic viewpoint.

A.3. What Happens to Errors in y When Nonlinear Models Are Linearized?

Linear least squares methods can be used for all linear models. Recall that the equations for parameters obtained from unweighted linear least squares are derived by assuming that there are no errors in the x_j and that the errors in the y_j(meas) have equal variances. This assumes explicitly that e_y has about the same value for a series of measured y_j, independent of the actual value of y_j. In the Beer's law example, we estimated a standard error in the absorbance of 0.003 units. This means that, on average, all the absorbance values in Tables 2.2 and 2.3 have $e_y = 0.003$.

The equal variance assumption that justifies the use of $w_j = 1$ can be verified experimentally. In the previous example, each y_j could be measured several times. From these results, a set of standard deviations s_j can be estimated, one for each y_j. If the s_j are approximately the same for all y_j, then the equal variance assumption holds.

There are other possibilities for the distribution of errors. For example, suppose that the errors in y_j are directly proportional to y_j. Intuitively, we need to place less weight on the results with larger errors. To achieve this, we minimize an error sum that contains a weighting factor $w_j = 1/y_j^2$. The correct form of the error sum (cf. eq. (2.9)) becomes

$$S = \sum_{j=1}^{n} [y_j(\text{meas}) - y_j(\text{calc})]^2/y_j^2. \tag{2.19}$$

The resulting equations for b_1 and b_2 will be different from those for the unweighted case. The new equations must be derived by using the expressions in eq. (2.10) with S as defined in eq. (2.19). This new set of simultaneous equations are solved for b_1 and b_2.

Another type of error distribution in y_j(meas) results for experiments using detectors that count events, such as photon or electron counters. In this case, errors in y_j depend on $y_j^{1/2}$, and $w_j = 1/y$. Yet another set of

equations for b_1 and b_2 results from this type of weighting. One common error in regression analysis involves using an unweighted regression when the distribution of errors requires a weighting factor.

In many cases it is possible to linearize a nonlinear model. However, it is necessary to consider how the linearization of a model changes the distribution of errors in y and x. We need to find out whether unweighted linear regression remains applicable after linearization [1]. As an example we consider the model for first-order decay, which is relevant to radioactive elements, chemical kinetics, luminescence, and many other situations [6]. The model is

$$y = y_o \exp(-kt) + b \qquad (2.20)$$

where y is the observed signal at time t, k is the rate constant for the decay, and y_o is the response at $t = 0$. The dependent variable y is a nonlinear function of the rate constant k, and b is a time invariant background parameter. If $b = 0$ or if b can be subtracted from y, we can linearize eq. (2.20) by taking the logarithm of both sides:

$$\ln(y - b) = -kt + \ln y_o. \qquad (2.21)$$

Linear regression with eq. (2.21) can now be undertaken, but we must realize that the new independent variable is not y but $\ln(y - b)$. We must consider the error distribution of this new independent variable in the linearized model. To follow this idea further, suppose we obtain data on the decay of a chemical species by using an absorbance spectrophotometer. We have seen that for such data, the constant variance assumption allows the use of $w_j = 1$. Suppose the constant error in y is $e_y = 0.001$ absorbance units. Because the error in y is constant, we can use it to compute the error in $\ln y$, assuming $b = 0$.

Table 2.4 shows the resulting error in the dependent variable $\ln y$ for three different y values. Here, $\ln y$ was computed from the value of y, and $e_{\ln y}$ was computed as $\ln(y + e_y) - \ln(y - e_y)$. We see in Table 2.4 that the error in $\ln y$ is not independent of $\ln y$. Even though e_y is independent of y, the error in the new independent variable $\ln y$ is not independent of

Table 2.4 Standard Errors in $\ln y$ for Different Values of y in Eq. (2.20)

y	e_y	$\ln y$	$e_{\ln y}$
1.000	±0.001	0	±0.001
0.100	±0.001	−2.303	±0.01
0.010	±0.001	−4.605	±0.1

its magnitude. The value of $e_{\ln y}$ increases as $\ln y$ becomes more negative. The assumption $w_j = 1$ is not justified for the linearized model in eq. (2.21). A rather complex weighting function is required for linear regression using this model.

Another complication is introduced if the background of the first-order decay experiment drifts with time. We can take this into account by adding a linear background term $mt + b$, where m is the slope of the background signal. Then the model becomes

$$y = A' \exp(-kt) + mt + b. \tag{2.22}$$

Taking the logarithm of both sides of this equation no longer provides a linear equation:

$$\ln y = -kt + \ln A' + \ln(mt + b). \tag{2.23}$$

Now, $\ln y$ depends on the logarithm of the background slope. On the other hand, *nonlinear regression* is directly applicable to models such as eq. (2.22).

Linear regression can also be used to analyze data with curvilinear models. Typical examples are polynomials in x; for example,

$$y = b_1 x^3 + b_2 x^2 + b_3 x + b_4 \tag{2.24}$$
$$y = b_1 / x^{1/2} + b_2 x^2 + b_3. \tag{2.25}$$

Because independent variable y is a linear function of the parameters $b_1 \ldots b_k$ in these models, linear regression can be used. Assuming that e_y is independent of y, eq. (2.9) is the correct form of the error sum S, and its derivatives with respect to $b_1 \ldots b_k$ are set equal to zero. The resulting simultaneous equations are solved for $b_1 \ldots b_k$ as usual. However, this procedure will result in a different set of equations for each curvilinear model.

We shall see in the next section that nonlinear regression is a convenient and easy to use general solution to all of these difficulties. It can be used to obtain direct fits of nonlinear models to experimental data with or without background components. These analyses do not require closed form equations for the parameters but depend on iterative numerical algorithms. Therefore, weighting can be approached in a general way without a new derivation for each separate problem. In principle, the same basic regression program can be used to fit any model.

B. Nonlinear Regression Analysis

B.1. Nonlinear Models

Single equation nonlinear models are those in which the dependent variable y depends in a nonlinear fashion on at least one of the parameters in the model. In general, nature is nonlinear. It is not surprising that nonlinear

models crop up in a wide variety of instrumental experiments used in chemistry and biochemistry. A few of these models are listed in Table 2.5. Note that in each of them y depends nonlinearly on at least one of the parameters $b_1 \ldots b_k$. Linear regression is not applicable to such models, but nonlinear regression can be applied generally to nonlinear models, to curvilinear models, and to models with no closed form representations.

As in linear regression, the goal of nonlinear regression is to find the absolute minimum in the error sum with respect to all the parameters. If we elaborate S in eq. (2.9) for a nonlinear model and set the first derivative with respect to each parameter equal to zero, we will find that the set of simultaneous equations does not yield closed form solutions. The alternative is to find the minimum S by numerical methods, using a sequence of repetitive mathematical operations called a *minimization algorithm*. This is what is done by nonlinear regression programs.

A mountain climbing analogy can be used to help understand how nonlinear regression works. When Sir Edmund Hillary climbed Mt. Everest, he and his team did not proceed to the summit in one long fast journey. They made a series of small journeys, one per day. At the end of each day, Sir Edmund and his colleagues reviewed their progress and planned the course of action for the next day's climb. In this way an efficient and successful assault on the summit was made, and the climbers reached the top of Mt. Everest.

Minimization algorithms used in nonlinear regression analysis can be viewed in a way similar to climbing a large mountain. We define a parameter space, with one axis for each parameter and an additional axis for the error sum S. For a specific model, this space includes an upside down mountain called an *error surface*. For a two parameter model, parameter space is three-dimensional. The axes represent the two parameters in the x-y plane, and S on the z axis (Figure 2.4).

The error surface describes the family of paths that the minimization algorithm may take to the minimum S. As in Sir Edmund's expedition, there can be a large number of possible paths. The minimum is approached

Table 2.5 Forms of Some Nonlinear Single-Equation Models

Models	Experiment
$y = b_1 \exp(-b_2 x)$	First-order decay kinetics
$y = b_1 \exp[-(x - b_2)^2/2b_3]$	Gaussian peak shape, NMR, IR, chromatography, etc.
$y = b_1/\{1 - \exp[(x - b_2)b_3]\}$	Sigmoid shape, polarography, steady state voltammetry [6]
$y = b_1/(1 + b_2^n x^{n-1}) + b_3 b_2^n x^{n-1}/(1 + b_2^n x^{n-1})$ $n \geq 2$	Thermodynamic linkage [7], e.g., diffusion of ligand–protein complex [5]

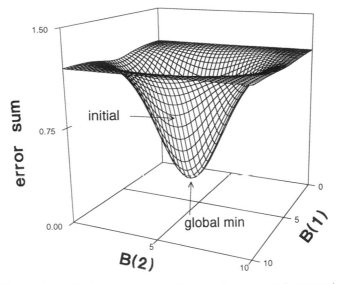

Figure 2.4   Idealized error surface for a nonlinear regression analysis.

in a series of steps, with each successive step planned on the basis of the previous one. Each step is designed to come a little closer to the minimum, until the absolute minimum is reached.

Computer programs for nonlinear regression analysis begin from a set of *best guesses* for the parameters in the model, as provided by the user. The minimization algorithm varies the parameters in a repetitive series of successive computational cycles, which are the steps leading toward the minimum S. At the end of each cycle, a new, and presumably smaller, value of S is computed from the new set of parameters found by the algorithm. These cycles are called *iterations*. A method based on successive computational cycles designed to approach the final answer is called an *iterative method*.

The goal of the minimization algorithm is to find the absolute minimum in the error sum S. Recall that the minimum S can be described by closed form equations for linear regression problems, but not for nonlinear models. The absolute minimum in the error sum S is the absolute or *global* minimum in the error surface.

The process of approaching and finding the minimum S is called *convergence*. This process can be envisioned graphically on an error surface in three-dimensional space (Figure 2.4). The initial guesses of the two parameters places the starting point of the calculation at an initial point on the error surface. The iterative minimization algorithm provides a stepwise journey along the error surface, which ends eventually at the global minimum.

In general, the coordinate system containing the error surface has $k +$ 1 dimensions, one for each of the k parameters and one for S. The approach to minimum S starting from a set of initial best guesses for the parameters can be viewed as the journey of a point toward the minimum of an error surface in a $(k + 1)$-dimensional orthogonal coordinate system. (In practice, the coordinates are not always fully orthogonal.)

Figure 2.4 represents an error surface for a model with two parameters, b_1 and b_2. Here $k = 2$, and the coordinate system is three-dimensional. The axis of the coordinate system in the Z direction corresponds to S and the other axes correspond to the parameters b_1 and b_2. The x and y axes are identified with these parameters. From initial guesses of b_1 and b_2, the initial point $P_1 [b_1, b_2, S_i]$ is located on the error surface. By systematic variation of the parameters, the algorithm employed by the program to minimize S causes this point to travel toward point $P_{min} [b_1, b_2, S_o]$. This is called the *point of convergence,* where S_o is at the absolute or *global minimum* on the error surface. The values b_1 and b_2 at S_o are the best values of these parameters with respect to the experimental data according to the least squares principle.

Reliable programs for nonlinear regression should contain criteria to automatically test for convergence limits. These limits are designed to tell the program when the minimum S has been reached. It is at this point in the computation that the best values of the parameters have been found. Here the program stops and displays the final parameter values and statistics and graphs concerning goodness of fit.

Convergence criteria are usually based on achieving a suitably small rate of change of S, of parameter values, or of statistics such as χ^2. For example, a program might terminate when the rate of change of S over 10 cycles is <0.02%. An alternative criteria might demand that the change in all parameter values is <0.005% on successive cycles. These criteria are included as logical statements in the regression programs and are usually tested at each cycle. Default values of the convergence limits suffice for most problems. In a few special cases, the user might want to change these limits. This is usually possible by simple changes in one or two lines of the program code.

B.2. Nonlinear Regression Algorithms

General programs that require the user to supply only the model and the data are available for nonlinear regression. Some of these are available in several commercial graphics software packages, and some have been written by independent authors [1, 6–8]. Many of these programs use numerical differentiation of the model so that analytical derivatives need not be provided by the user.

Commercial software packages for mathematics such as Mathcad and Matlab have built in minimization functions that allow facile construction of general programs for regression analysis. A version of a nonlinear regression program in the Mathcad environment is provided in Section B.4 of this chapter.

Programs for nonlinear regression differ mainly in the algorithm used to minimize the error sum S. General programs have a "model subroutine" into which the user writes the program code for the desired regression model. The model can be an explicit closed form equation, a series of equations, or a numerical simulation, as long as it supplies computed values of the dependent variable y_j(calc) to the main program.

A detailed discussion of nonlinear regression algorithms falls outside the aims of this book. In the following, we give a qualitative description of some of the more popular algorithms.

One commonly used algorithm employs the *steepest descent* method [4], in which the search for the minimum S travels on the gradient of the error surface. The gradient is estimated numerically by finite difference methods. The progress toward the minimum S is monitored at each iteration or cycle and adjusted when necessary. Steepest descent provides fast convergence in the initial stages of the computation, but slows down considerably near the convergence point. Such programs written with conservative tolerances for convergence are extremely reliable, but may be very slow for some applications.

The *Gauss–Newton* algorithm approximates nonlinear models for y_j(calc) by linear Taylor series expansions [4]. After initial guesses, new values of the parameters are found at each cycle by methods similar to linear least squares, using expressions involving first derivatives of S with respect to each parameter. Ideally, each iterative cycle gives successively better estimates for the parameters until the absolute minimum in S is reached. Unlike the steepest descent method, convergence is often fast in the vicinity of the minimum.

The *Marquardt–Levenberg* algorithm [5] contains elements of both steepest descent and Gauss–Newton methods but converges more rapidly than either of these. The Marquardt–Levenberg algorithm behaves like an efficient steepest descent method when the parameters place the error sum far from minimum S on the error surface. Close to the minimum S, it behaves like the Gauss–Newton method, again under conditions where the latter algorithm is efficient.

Another algorithm that can be used for nonlinear regression is the *modified simplex*. For k parameters, this method employs a $(k + 1)$-dimensional geometric construct called a *simplex*. A three-dimensional simplex is a triangle. A set of rules are used by the algorithm to move the simplex toward the minimum of the error surface as it simultaneously contracts in size [9]. Convergence times are comparable to steepest descent methods but longer than the Marquardt–Levenberg method.

Nonlinear regression programs can be interfaced with graphics subroutines that plot the experimental data along with the curve computed from the model. If such a plot is consulted after the initial parameter guesses are made, but before the iterations begin, a serious mismatch of experimental and computed curves can immediately be recognized and better initial guesses can be made before the regression is started. The graphics subroutine can also be used as a rough visual check of final goodness of fit, as discussed already.

Nonlinear regression programs should include criteria to automatically test for convergence and terminate the program when preset conditions for convergence are reached. As mentioned previously, reliable tests for convergence are often based on the rate of change of S or on the rate of change of the parameters. Criteria for convergence can often be adjusted by the user, but this is normally not necessary. Although convergence can be tested over a series of cycles, this is self-defeating in terms of computational time in an algorithm as fast as Marquardt–Levenberg, and convergence is tested on successive cycles.

The job of regression algorithms is to find the global minimum of S. If the error surface is irregularly shaped, the algorithm could conceivably stop at a *local minimum* with an error sum considerably larger than the global minimum. Parameter values would then be in error. Although such *false minima* are possible, they are rarely encountered in practice in well-conditioned, properly designed regression analyses. A more complete discussion of error surfaces is included in Chapter 4.

A graphic view of the initial computed response in regression analysis aids in avoiding starting points too far from the absolute minimum, which might cause convergence to a false minimum. Local minima can be detected by starting regression analyses with several different, but physically reasonable, sets of initial values of the parameters. Convergence to significantly different values of S and different final parameters for different starting points suggests that local minima are being reached. Convergence to identical values of S and identical parameters for any reasonable starting point indicates that the global minimum has been found.

B.3. Matrix Representation for Nonlinear Regression

A general single-equation nonlinear model may be expressed in matrix notation as [6]

$$\mathbf{Y}_c = [\mathbf{F}(\mathbf{x}_j)]'\mathbf{b} \qquad (2.26)$$

where $\mathbf{Y}_c$ is the model vector containing the computed $y_j(\text{calc})$, $[\mathbf{F}(\mathbf{x}_j)]'$ is the transpose of a column matrix expressing the functionality of the model, and $\mathbf{b}$ is the vector containing the values of the model parameters. If $\mathbf{Y}$ is the matrix containing the values of y_j(meas), the unweighted error sum S becomes

$$S = [\mathbf{Y} - \mathbf{Y_c}]'[\mathbf{Y} - \mathbf{Y_c}].$$ (2.27)

A detailed discussion of parameter estimation by matrix methods can be found in the book by Bates and Watts [5]. Bevington's book [3] provides bridges between the algebraic and matrix approaches.

B.4. Example Regression Analysis for an Exponential Decay

Problem. The fluorescence intensity of a reactant is measured to obtain its concentration in millimoles vs. time for a suspected first-order decomposition reaction. The data are read from a file in the form of y = reactant concentration in mM and x = time in seconds. Assuming that errors in y are independent of y (i.e., errors are *absolute*), use nonlinear regression to see if first-order decay holds and to estimate the rate constant and the pre-exponential factor.

This example demonstrates how nonlinear regression analysis is used. The data were analyzed by using a program for nonlinear regression in the Mathcad environment. The program, data, graphs, and results are integrated together in this format, as shown in Box 2.2. We considered time to be free of error. Therefore, only the random errors in y are significant. Because errors in y are absolute, we minimize S in eq. (2.9) with $w_j = 1$. The model is given in eq. (2.20),

$$y = y_o \exp(-kt) + b$$ (2.20)

where b is assumed to be zero. There are two parameters: $b_1 = y_o$ and $b_2 = k$, the first-order rate constant in s^{-1}. These data were calculated from eq. (2.20) and normally distributed random noise was added to y. Thus, we know that the true values are $b_1 = 14.0$ mM and $b_2 = 15.0$ s^{-1}.

We now describe the elements of the Mathcad calculation. The first equation reads the data from a file called CURDAT1. Data in Mathcad are read from a data file as a matrix. The following equations display this data matrix as M and define the data vectors Y and x for the program. These data vectors are $n \times 1$ matrices, also called *column vectors*. Y contains all the y_j and x contains all the x_j, respectively.

The number of data pairs (n) to be analyzed must be provided in the next section. Next, the initial values of the parameters b_1 and b_2 are guessed. These should be best guesses based on inspection of the data and can be adjusted by the user. These best guesses define the starting point of the regression analysis. The model function to be fit to the data, $F(x, b_2, b_2)$, is provided in the next section.

The first graph is the result of a calculation of the model $F(x_j, b_1, b_2)$ with the initial parameters, given by the dotted line, plotted together with

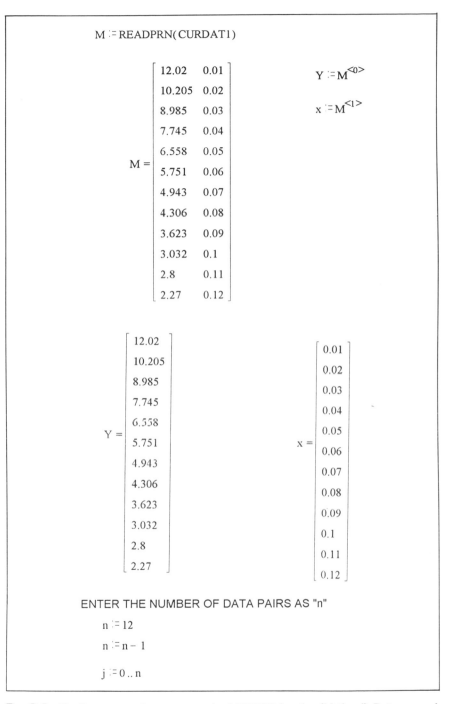

$$M := READPRN(CURDAT1)$$

$$M = \begin{bmatrix} 12.02 & 0.01 \\ 10.205 & 0.02 \\ 8.985 & 0.03 \\ 7.745 & 0.04 \\ 6.558 & 0.05 \\ 5.751 & 0.06 \\ 4.943 & 0.07 \\ 4.306 & 0.08 \\ 3.623 & 0.09 \\ 3.032 & 0.1 \\ 2.8 & 0.11 \\ 2.27 & 0.12 \end{bmatrix}$$

$$Y := M^{<0>}$$

$$x := M^{<1>}$$

$$Y = \begin{bmatrix} 12.02 \\ 10.205 \\ 8.985 \\ 7.745 \\ 6.558 \\ 5.751 \\ 4.943 \\ 4.306 \\ 3.623 \\ 3.032 \\ 2.8 \\ 2.27 \end{bmatrix} \qquad x = \begin{bmatrix} 0.01 \\ 0.02 \\ 0.03 \\ 0.04 \\ 0.05 \\ 0.06 \\ 0.07 \\ 0.08 \\ 0.09 \\ 0.1 \\ 0.11 \\ 0.12 \end{bmatrix}$$

ENTER THE NUMBER OF DATA PAIRS AS "n"

$$n := 12$$

$$n := n - 1$$

$$j := 0 .. n$$

Box 2.2 Nonlinear regression program using MINERR function (Mathcad). Data are read from an ASCII file as x,y pairs.

b1 := 16   b2 := 28

Function to be fit to data: exponential decay (change F to any function you wish to fit)

$F(x, b1, b2) := b1 \cdot \exp(-b2 \cdot x)$

GRAPH OF DATA + MODEL FOR INITIAL PARAMETERS

$$SSE(b1, b2) := \sum_j (Y_j - F(x_j, b1, b2))^2$$

Given

$SSE(b1, b2) = 0$ $1 = 1$

$\begin{pmatrix} b1 \\ b2 \end{pmatrix} := Minerr(b1, b2)$

Values of the parameters:

Sum of squares:

b1 = 13.907

b2 = 14.846 $SSE(b1, b2) = 0.0659$

Root mean square error:

$$\sqrt{\frac{SSE(b1, b2)}{n - 2}} = 0.0856$$

Box 2.2 *(continues)*

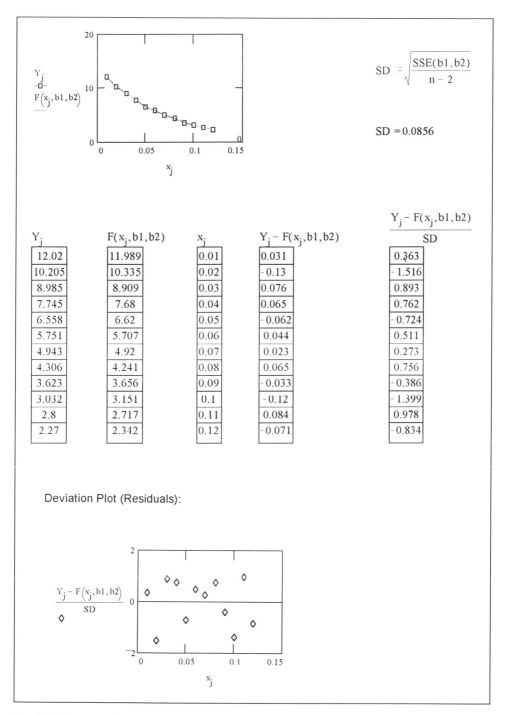

$$SD := \sqrt{\frac{SSE(b1,b2)}{n-2}}$$

$$SD = 0.0856$$

Y_j	$F(x_j,b1,b2)$	x_j	$Y_j - F(x_j,b1,b2)$	$\dfrac{Y_j - F(x_j,b1,b2)}{SD}$
12.02	11.989	0.01	0.031	0.363
10.205	10.335	0.02	-0.13	-1.516
8.985	8.909	0.03	0.076	0.893
7.745	7.68	0.04	0.065	0.762
6.558	6.62	0.05	-0.062	-0.724
5.751	5.707	0.06	0.044	0.511
4.943	4.92	0.07	0.023	0.273
4.306	4.241	0.08	0.065	0.756
3.623	3.656	0.09	-0.033	-0.386
3.032	3.151	0.1	-0.12	-1.399
2.8	2.717	0.11	0.084	0.978
2.27	2.342	0.12	-0.071	-0.834

Deviation Plot (Residuals):

Box 2.2 (*continued*)

the data points, y_j. If agreement between the points and the line is not way out of line, the user can begin the regression by telling the program to calculate the Mathcad document. (An *Automatic Calculate* mode can also be chosen.) In many cases, sufficient initial agreement may require simply that both plots fall in the same coordinate space, as shown in this example. If agreement is unsatisfactory, a new set of initial parameters can be chosen to give a better initial agreement.

The next line in the program defines the error sum as the sum of squares of the errors, $SSE(b_1, b_2)$. The several lines following use the MINERR function of Mathcad, which employs the Marquardt–Levenberg algorithm to find the minimum in the error sum. (The reader is directed to the Mathcad manual for more details. See the end of this chapter for the source.) The remainder of the Mathcad program gives the final results, including parameter values, minimum error sum (sum of squares), and root mean square error. In cases such as this one, where the errors are absolute, the root mean square error is the same as the standard deviation of the regression.

The second graph in the document is a plot of the data along with the model computed with the final parameters. This is followed by lists of all input and computed data, and a final plot of the residuals; that is, $[y_j(\text{meas}) - y_j(\text{calc})]/\text{SD}$ on the vertical axis plotted against the independent variable on the horizontal axis [1].

The quantities $[y_j(\text{meas}) - y_j(\text{calc})]/\text{SD}$ are sometimes given the symbol dev_j. The dev_j are the differences of each experimental data point from the calculated regression line divided by the standard deviation (SD) of the regression. Since the input data had randomly distributed errors, the residual plot shows a random scatter of points about the zero residual axis; that is, $\text{dev}_j = 0$. As for linear regression, this type of plot is evidence that the model provides an acceptable fit to the data.

In our example, we chose a value of b_2 rather far from the final convergence value, but the nonlinear regression analysis converged rapidly to the values of the parameters very close to the true values of $b_1 = 14$ mM and $b_1 = 15$ s^{-1}. The errors in the parameters are related to the random error added to the data. The standard deviation of the regression (SD) is about 0.7% of the largest y value. This is slightly less than the amount of random error added to the data, which was 1% of the largest y value. This is another indicator of an excellent fit; that is, $\text{SD} \leq e_y$, as discussed for linear regression. For real experimental data, e_y is the estimate of random error in the measured signal.

Therefore, we see that the *goodness of fit* criteria for nonlinear regression analyses are similar to those for linear regression. Ideally, the residual plot should be random, and we should be able to achieve $\text{SD} \leq e_y$ for a good fit. The example also illustrates another important practice, that of testing an unfamiliar analysis model with theoretical data having known parameters before proceeding to analysis of experimental data. This concept will be illustrated further in Chapter 3.

The Mathcad program used in the preceding example is general and can be used with any model. The user needs to enter only the model $F(x, b_1, \ldots, b_k)$, to provide the initial guesses for the parameters, and indicate a suitable data file. Note that the structure of the data file is pairs of y_j, x_j, with one data point pair per line as in a typical ASCII file.

B.5. Weighting

As in linear regression, the relation between random errors in experimental values of dependent and independent variables dictates the choice of weighting factors w_j in the error sum in eq. (2.9). For a model with the general form

$$y_j = F(x_j, b_1, \ldots, b_k) \tag{2.28}$$

standard errors in both the dependent variable y and the independent variable x may be significant. However, a useful simplifying assumption that we have used thus far is that standard error in x is negligible. That is, the x_j are known with absolute precision. This commonly used assumption is reasonably accurate in many cases, such as when y is a slowly varying instrumental response measured vs. time as x. If random errors in x can be neglected, only the random errors in y need be considered in the error sum and the weighting factor.

In the example of Section B.4, the error in y was independent of the size of y and the unweighted error sum was minimized. In such cases, we say that there is *absolute* error in y. Suppose we measure the optical absorbance of a reacting species vs. time with a conventional spectrophotometer. As long as the sensitivity scale is the same for all measurements, the standard error in absorbance (y) is likely to be the same whether the absorbance is small or large. Therefore, we have *absolute* error in y. This can be checked experimentally by measuring standard deviations of the absorbance. If the errors are absolute, the standard deviations should be approximately the same at different absorbance values in the range of the experimental data.

In other cases, standard errors in y may be proportional to y, and the sum of squares of the *relative* errors should be minimized. Standard errors in y may be neither absolute nor strictly relative, and alternative weighting factors are required. This occurs for data obtained by counting detectors, for which the random error in y(meas) is Poisson distributed [10]; that is, proportional to $[y(\text{meas})]^{1/2}$. Appropriate weighting factors for these common cases are listed in Table 2.6.

Table 2.6   Common Weighting Factors for Errors in y_j Assuming Error-Free x_j

Error distribution	Weighting, w_j (eq. (2.9))	SD[a]
Absolute, independent of y_j	1	$SD = \sqrt{\dfrac{[y_j(\text{meas}) - y_j(\text{calc})]^2}{(n-p)}}$
Relative, proportional to y_j	$[y_j(\text{meas})]^{-2}$	$SD = \sqrt{\dfrac{[y_j(\text{meas}) - y_j(\text{calc})]^2}{(n-p)\,y_j(\text{meas})^2}}$
Proportional to $y_j^{1/2}$	$[y_j(\text{meas})]^{-1}$	$SD = \sqrt{\dfrac{[y_j(\text{meas}) - y_j(\text{calc})]^2}{(n-p)\,y_j(\text{meas})}}$

[a] n = number of data; p = number of parameters

The use of weighting factors in the Mathcad nonlinear regression program simply requires its inclusion in the expression in S as $SSE(b_1, b_2)$, using the form in eq. (2.9). In general, for a Poisson error distribution in y,

$$S = \sum_{j=1}^{n} [1/y_j(\text{meas})][y_j(\text{meas}) - y_j(\text{calc})]^2. \qquad (2.29)$$

For weighted regression however, the SD has a different relation to the root mean square error. These are also listed in Table 2.6.

Finally, eq. (2.29) expressed the sum of squares of the differences in calculated and experimental values of y. If random errors in x are also significant, an appropriately weighted function reflecting the distribution of errors in both x and y should be minimized [6].

C. Sources of Mathematics Software Capable of Linear and Nonlinear Regression

1. Mathcad, mathematical software package available for IBM PC and Macintosh-type computers from Mathsoft, Inc., 201 Broadway, Cambridge, MA 02139. The program contains a Marquardt–Levenberg minimization algorithm. An advantage of this software is that it presents equations just as you would write them or see them in a mathematics textbook. It is somewhat easier to learn than Matlab, but less powerful.

2. Matlab, high-performance numerical computation and visualization software, available from The Mathworks, Inc., 24 Prime Park Way, Natick, MA 01760. The program contains a simplex minimization algorithm in the main package and Marquardt–Levenberg and Gauss–Newton algorithms in the Optimization Toolbox, supplied separate from the main package.

References

1. L. Meites, "Some New Techniques for the Analysis and Interpretation of Chemical Data," *Critical Reviews in Analytical Chemistry* **8** (1979), pp. 1–53.
2. J. C. Miller and J. N. Miller, *Statistics for Analytical Chemistry,* 2nd Ed. Chichester, England: Ellis Horwood, 1988.
3. P. R. Bevington, *Data Reduction and Error Analysis for the Physical Sciences.* New York: McGraw-Hill, 1969.
4. J. H. Kalivas, in S. J. Haswell (ed.), *Practical Guide to Chemometrics,* pp. 99–149. New York: Marcel Dekker, 1992.
5. D. M. Bates and D. G. Watts, *Nonlinear Regression Analysis and Its Applications,* pp. 1–30. New York: Wiley, 1988.
6. J. F. Rusling, "Analysis of Chemical Data by Computer Modeling," *CRC Critical Reviews in Analytical Chemistry* **21** (1989), pp. 49–81.
7. T. F. Kumosinski, "Thermodynamic Linkage and Nonlinear Regression Analysis: A Molecular Basis for Modeling Biomacromolecular Processes," *Advances in Food Science and Nutrition* **34** (1990), pp. 299–385.
8. J. F. Rusling, in S. J. Haswell (ed.), *Practical Guide to Chemometrics,* pp. 151–179. New York: Marcel Dekker, 1992.
9. S. N. Deming and S. L. Morgan, *Anal. Chem.* **45** (1973), p. 278A.
10. K. J. Johnson, *Numerical Methods in Chemistry,* pp. 282–303. New York: Marcel Dekker, 1980.

Chapter 3

Building Models for Experimental Data

A. Sources of Data and Background Contributions

We can envision at least two different sources of data that could be analyzed by nonlinear regression. One type of data is a typical *response curve* from a single instrumental experiment. This curve can be represented as a series of digital data points obtained as the response measured by the instrument vs. time or vs. an experimentally controlled variable. Some familiar examples are given in Table 3.1. The measured response is the dependent variable y_j(meas). The respective independent variables x_j, are listed in Table 3.1.

A second type of data could involve a physical quantity derived from a series of experiments vs. a series of values of an experimentally controlled variable, which is different in each experiment. In this case, one data point results from each experiment. Some examples might be the absorbance at a fixed wavelength vs. the concentration of absorber in a series of different solutions or measured diffusion coefficients in solutions of an aggregating protein vs. the concentration of the protein.

We distinguish between these two types of data because it is necessary to account for the background in analyses of the data. Either the regression models themselves, or data manipulation prior to regression analyses, is required to account for all contributions to the measured signal. Thus, a generic expression for a single equation model (cf. Chapter 2, Section B.1) is

$$y = \text{signal of interest} + \text{background}. \tag{3.1}$$

For example, for absorbance vs. concentration data, the background might take the simple form of a constant offset in y; that is, a blank value

Table 3.1 Variables for Some Typical Response Curves

Experiment	Dependent variable	Independent variable
UV-Vis Spectroscopy	Absorbance	Wavelength
Voltammetry	Current	Applied voltage
Fluorescence Spectroscopy	Signal intensity	Wavelength or energy
Fluorescence decay	Signal intensity	Time
Photoelectron spectroscopy	Detector counts	Energy
Chromatography	Detector signal	Time
Mass spectrometry	Relative intensity	Mass/charge
FT-IR spectroscopy	Absorbance	Wave number

of the absorbance in a solution of zero concentration. The diffusion coefficient data mentioned previously, if properly corrected for artifacts of the measurement, may have a zero background term. On the other hand, a full instrumental response curve may contain background drift, nonrandom noise, and other instrumental bias added to the signals from the chemical or physical events being studied. Thus, the background term in eq. (3.1) may depend on x. Instrumental contributions may include finite background offset, drift, instability, and other characteristics of the measuring system that yield a finite signal not characteristic of the sample. We will refer to these instrumental signatures collectively as *background*.

If background contributions to an instrumental response curve are reasonably constant with time or if their variations with time are highly reproducible, they can sometimes be subtracted from the data. A response curve for a blank not containing the analyte might be recorded separately and subtracted from the response curve of the sample. However, situations in which this approach is successful are limited and require a highly reproducible background that is identical for the blank and the sample.

An alternative to subtraction is to include specific terms in the model accounting for the background. This approach assumes that the background can be described accurately in mathematical form. In the best cases, where signal to noise ratio is large and the background drifts slowly, a simple linear background term may be successful [1, 2].

Suppose a measured exponential decay response has a background that is constant with time. An appropriate model is

$$y_j = b_1 \exp(-b_2 x_j) + b_3. \tag{3.2}$$

Consider an example set of exponential decay data similar to that in Section 2.B.4, but with a constant background term b_3 that is about 25% of the maximum y_j. A fit of such data was done to the model without background:

$$y_j = b_1 \exp(-b_2 x_j). \tag{3.3}$$

Table 3.2 Results of the Two-Parameter Model in Eq. (3.3) Fit onto Exponential Decay Data[a]

Parameter/statistic	Initial value	True value	Final value ($b_3 = 0$)	Final value ($b_3 = 3$)
b_1	16	14	13.91	16.32
b_2	28	15	14.85	10.04
SD		($e_y = 0.15$)	0.085	0.246
Deviation plot			Random	Nonrandom

[a] Data generated by using Eq. (3.2) with absolute random noise at 1% of the maximum y. One set of data was generated with $b_3 = 3$, and the other with $b_3 = 0$.

The results from this analysis by a Marquardt nonlinear regression program are summarized by Table 3.2. The data are the same as those discussed in Section 2.B.4, but we have included a second data set with a constant background added. As seen previously, eq. (3.3) fit the data without any background ($b_3 = 0$) quite well. Final parameter values were close to the true values, SD $< e_y$ (Table 3.2), and the deviation plot had a random scatter of points.

On the other hand, the fit of the data with a constant background $b_3 = 3$ shows final parameter values that are in error by as much as 33% when compared to the true values (Table 3.2). A value of SD $< e_y$ and a clearly nonrandom deviation plot (Figure 3.1) are clear indicators that the model provides a poor fit to the data. Because we generated the data with eq. (3.2), we can surmise that the reason for this poor fit is that the background b_3 is not included in the model. This simple example illustrates the fact that very large errors can result from the neglect of background in nonlinear regression.

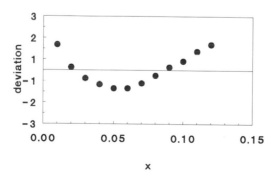

Figure 3.1 Nonrandom deviation plot corresponding to results in Table 3.2 for nonlinear regression of exponential data with a constant background to a model not considering background.

Table 3.3 Results of the Three-Parameter Model in Eq. (3.2) Fit onto Exponential Decay Data with a Constant Background

Parameter/statistic	Initial value	True value[a]	Final value ($b_3 = 3$)
b_1	16	14	13.97
b_2	28	15	14.89
b_3	2.5	3	2.99
SD		($e_y = 0.15$)	0.0224
Deviation plot			Random

[a] Data generated by using eq. (3.2) with absolute random noise at 1% of the maximum y.

The correct model (eq. (3.2)) fits the decay data with $b_3 = 3$ with excellent results. This analysis (Table 3.3) gives final parameter values with small errors, a random deviation plot (Figure 3.2), and a SD that is 10-fold smaller than the fit to eq. (3.3).

Therefore, the fit of the model including the background term is successful. The value of SD is 10-fold smaller after the addition of the third parameter to the model. The goodness of fit of two and three parameter models such as eqs. (3.3) and (3.2) can also be compared by using the extra sum of squares F test [1], which is discussed in more detail later in this chapter (Section C.1).

In the preceding example, we were lucky enough to be able to model the background in the exponential decay with a simple constant offset term. In general, the background may depend upon the independent variable. For example, the measured background may vary with time. In some cases involving approximately linear drift, a linear background term may suffice. In other situations, more complex expressions for background may be necessary. Some useful background expressions that can be added to regression models are listed in Table 3.4.

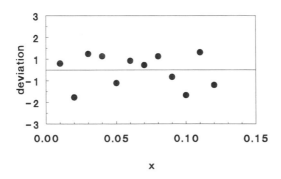

Figure 3.2 Random deviation plot corresponding to results in Table 3.3 for nonlinear regression of exponential data with a constant background onto the correct model.

Table 3.4 Common Background Expressions for Regression Models

Background model	Shape	Possible application
b_3	Constant offset	Constant blank
$b_3 \exp(\pm b_4 x)$	Increasing or decreasing exponential	Voltammetry for peaks near the ends of potential window
$b_3 x + b_4$	Linear	Drifting baseline
$b_2/\{1 - \exp[(\pm x - b_2)b_3]\}$	Sigmoidal variation	X-ray photoelectron spectroscopy

In our experience, there seems to be no general way to predict how to account for background in every experiment. Each experimental situation is a little bit different. We recommend establishing the best way to deal with background well before analyzing the real experimental data. If the form of the background is well known, this can be approached initially by tests with computer generated data, as already illustrated. The data generated should contain added noise to eliminate the effect of computer round-off errors in final statistics and deviation plots and to realistically mimic experimental conditions.

A BASIC subroutine that can be used to add randomly distributed noise to data is described in Box 3.1. Random number generators on computers

```
SUBROUTINE 3.1. BASIC SUBROUTINE FOR ADDITION OF RANDOM
                           NOISE TO DATA

10 REM GENERATES A SET OF RANDOM NUMBERS WITH ZERO MEAN AND UNIT
VARIANCE
20 REM these numbers can be used to generate noise in data
30 REM generate your data then add noise by adding to each point a value
40 REM = (fraction of noise desired)*ZZ(I%)*(max value of y) for absolute
noise for N1% data points
50 REM This program may be used as a subroutine in the program you use
to generate data
60 REM NAME - RANDOM.BAS
70 INPUT "NO. OF DATA NEEDED "; N1%
80 DIM X1(N1%), ZZ(N1%)
90 RANDOMIZE VAL(RIGHT$(TIME$,2))
100 FOR J% = 1 TO N1%
110 X1(J%) = RND: PRINT X1(J%)
120 NEXT J%
130 PRINT "DATA POINT #","RANDOM #","DATA POINT #","RANDOM #"
140 FOR I% = 1 TO N1%/2
150  ZZ(2*I%-1)=(-2*LOG(X1(2*I%-
1))/LOG(10))^(1/2)*COS(6.283*X1(2*I%))
160  ZZ(2*I%)=(-2*LOG(X1(2*I%-1))/LOG(10))^(1/2)*SIN(6.283*X1(2*I%))
170  PRINT 2*I%-1,ZZ(2*I%-1),2*I%,ZZ(2*I%)
180 NEXT I%
190 RETURN
```

Box 3.1 BASIC subroutine for addition of random noise to data.

do not give a normally distributed set of numbers. The subroutine here uses the method described by Box and Muller [3] to convert random numbers generated by the BASIC RND function to N normally distributed numbers z, with a mean of 0 and a variance of 1. This set of numbers is used to add noise to a set of values of y.

Once an initial survey with noisy theoretical data is completed, it is important to base the final decision on how to deal with background on real experimental data. If possible, the testing should be done by using data for a *standard system* with well-known parameters. For example, in the exponential decay case we might choose as our standard a chemical that decays with a well-known rate constant. We would have the most confidence in the method of background treatment that returns the values of the parameters with the least error under experimental conditions similar to those to be used in studies of our unknown systems. This procedure follows from common practice in the physical sciences. If we are developing a new method, we first test it with a reference standard.

An auxiliary approach may be used if representative background response curves can be obtained in the absence of the sample. For example, in the exponential decay experiment, one could measure the signal vs. time and fit these data to various background models. The model that best fits the background data can then be added to the model for the sample data. However, it is not always possible to obtain such background data. It must be ensured that the presence of the sample does not change the background.

B. Examples of Model Types

B.1. Closed Form Theoretical Models

We can generalize from the preceding discussion that a nonlinear regression model must contain a mathematical description of the signal resulting from the sample in addition to a description of the background, unless the background can be accurately subtracted from the raw data. Theoretical equations describing the instrumental response are excellent starting points for such regression models. When used with the appropriate background terms, computations are usually fast, and physically significant parameters are obtained. In this section, we present another example of such a model. The first such example was the exponential decay system presented in the previous chapter.

Response curves of current (i) in an electrolytic cell containing a planar working electrode can be obtained when the potential (E) is varied linearly with time. Under experimental conditions, where a steady state exists between the rate of electrolysis and the rate of mass transport to the electrode, sigmoid-shaped i–E curves (Figure 3.3) result. These conditions can be

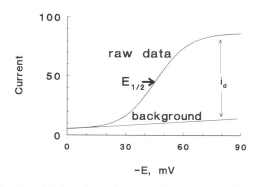

Figure 3.3 Sigmoid-shaped steady state voltammogram with background.

achieved by stirring the solution, by rotating the electrode, or by using very tiny electrodes. These sigmoid curve shapes are therefore found in the electrochemical techniques of polarography, normal pulse voltammetry, rotating disk voltammetry, and slow scan voltammetry using disk "ultramicroelectrodes" with radii less than about 20 μm. The results of such experiments are called *steady state voltammograms*.

The simplest electrode process giving rise to a sigmoid-shaped steady state response is a single-electron transfer unaccompanied by chemical reactions. Thus, O the oxidant, and R the reductant, are stable on the experimental time scale.

$$O + e \rightleftharpoons R \quad (E^{0\prime}). \tag{3.4}$$

Equation (3.4) represents a fast, reversible transfer of one electron to an electroactive species O from an electrode. The resulting voltammetric current-potential (i vs. E) curve centers around the formal potential $E^{0\prime}$ (Figure 3.3). The half-wave potential ($E_{1/2}$) is the value of E when the current is one half of the current at the plateau between about -70 to -90 mV. This plateau current is called the *limiting* or *diffusion current* (i_d) and is proportional to the concentration of O in the solution. $E^{0\prime}$ is the same as the half-wave potential if the diffusion coefficients of O and R are equal.

Assuming that the solution contains only O initially in the solution, the equation describing the steady state faradaic i vs. E response for eq. (3.4) is

$$i = i_1/(1 + \theta) \tag{3.5}$$

where $\theta = \exp[(E - E^{0\prime})(F/RT)]$, R is the universal gas constant, F is Faraday's constant, and T is the temperature in Kelvins. The use of this model for nonlinear regression provides estimates of $E^{0\prime}$ and i_1. A typical example for data without background shows excellent results (Table 3.5) when evaluated by our usual criteria.

Table 3.5 Results of the Model in Eq. (3.5) Fit onto Steady State Voltammetry Data

Parameter/statistic	Initial value	True value[a]	Final value
$b_1 = i_1$	1.01	1.000	1.003
$b_2 = F/RT$	36	38.92	38.72
$b_3 = E^{0\prime}$	−0.25	−0.200	−0.2000
SD		($e_y = 0.005$)	0.0028
Deviation plot			Random

[a] Data generated by using eq. (3.5) with absolute normally distributed noise at 0.5% of the maximum y.

For a well-resolved experimental steady state reduction curve with signal to noise ratio of 200/1 or better in the limiting current region, an appropriate model [1] for a regression analysis combines eq. (3.5) with a linearly varying background current:

$$y = b_1/(1 + \theta) + b_4 x + b_5. \qquad (3.6)$$

The $b_1/(1 + \theta)$ term on the right-hand side of eq. (3.6) describes the faradaic current from electron transfer in eq. (3.4) and also contains the parameters $b_2 = F/RT$, and $b_3 = E^{0\prime}$. The value of b_2 is known and fixed by the value of T. It is included as an adjustable parameter in the model as an additional check for correctness. Its value returned by the regression program can be checked against the theoretical value.

The $b_4 x + b_5$ terms on the right-hand side of eq. (3.6) accounts for the background. The parameter b_4 is the slope of the background current, in $\mu A/V$, for example, and b_5 is the current intercept in μA at zero volts ($x = 0$). A correction factor accounting for observed nonparallel plateau and baseline is sometimes included in the model but may not always be necessary [1]. The quantities b_4 and b_5 may be used as regression parameters, or one or both of them can be kept fixed. In the latter case, their values must be measured from data obtained in a range of potentials before the rise of the sigmoid wave and then provided to the program.

As a second example of the importance of background considerations, we generated a steady state voltammogram by using eq. (3.6) and fit these data to eq. (3.5), which has no background terms. With a rather small linear background having a slope of 0.2 $\mu A/V$, errors of about 10% are found in parameters b_1 and b_2 (Table 3.6). The value of SD is about fivefold larger than that found for data without background (Table 3.5), and SD $> e_y$, where e_y here is the amount of absolute normally distributed noise added to the data, 0.005. The data points fall away from the regression line, and the deviation plot is clearly nonrandom (Figure 3.4).

If you have read the preceding pages of this book, you already know

Table 3.6 Results of the Model in Eq. (3.5) without Background Term Fit onto Steady
State Voltammetry Data with Background

Parameter/statistic	Initial value	True value[a]	Final value	Error
b_1	1.01	1.000	1.084	8.4%
b_2	36.0	38.92	35.03	10.0%
b_3	−0.21	−0.200	−0.198	1%
b_4	Not used	−0.200		
b_5	Not used	0.005		
SD		($e_y = 0.005$)	0.0107	
Deviation plot			Nonrandom	

[a] Data generated by using eq. (3.5) with absolute normally distributed noise at 0.5% of
the maximum y.

how to correct the errors in this example. We can simply fit the data to
eq. (3.6), which contains the linear background expression. The results of
such an analysis (Table 3.7) indicate a good fit of the model to these data.
The errors in the parameters and SD are four- to fivefold smaller than in

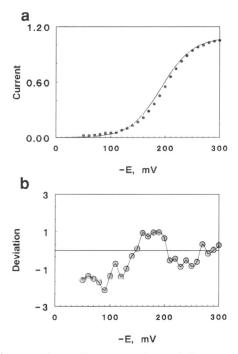

Figure 3.4 Graphic output for nonlinear regression analysis summarized in Table 3.6 for
steady state voltammetry data with a constant background fit using a model not considering
background: (a) points are experimental data, line computed from regression analysis;
(b) nonrandom deviation plot.

Table 3.7 Results of the Model in Eq. (3.6) with Background Term Fit onto Steady State Voltammetry Data with Background

Parameter/statistic	Initial value	True value[a]	Final value	Error
b_1	1.01	1.000	0.983	1.7%
b_2	36.0	38.92	39.68	2.0%
b_3	−0.21	−0.200	−0.1996	0.2%
b_4	−0.22	−0.200	−0.267	34%
b_5	0.01	0.005	0.0051	2%
SD		($e_y = 0.005$)	0.0033	
Deviation plot			Random	

[a] Data generated by using eq. (3.6) with absolute normally distributed noise at 0.5% of the maximum y.

the fit without background parameters (cf. Table 3.6). The data points all fall on the regression line (Figure 3.5), the deviation plot is random, and SD $< e_y$. The error in the background slope b_4 is relatively large because its value is small and it contributes little to the total signal. Information

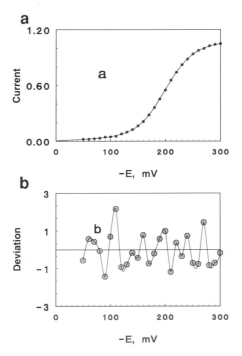

Figure 3.5 Graphic output for nonlinear regression analysis summarized in Table 3.7 for steady state voltammetry data with a constant background fit using the correct model: (a) points are experimental data, line computed from regression analysis; (b) random deviation plot.

about this parameter is not well represented in the data. Also, b_4 is partly correlated with b_1. Partial correlation between parameters can cause an increase in their errors. Effects of correlation, that is, the dependence of one parameter in the model on another parameter, will be discussed in more detail later in this chapter.

The model in eq. (3.6) for steady state voltammetry was constructed by adding a background term to a theoretical expression describing the faradaic response caused by the electrode reaction. In this example, we employed a linear background, but the same general conclusions would be drawn from data with any mathematical form of the background. We shall see in the applications chapters that adding a background term to a theoretical expression for the sample response is a useful approach to model building in a variety of situations.

B.2. Peak-Shaped Data

Many experiments in chemistry and biochemistry give peak-shaped signals. For methods such as chromatography, overlapped peaks result from poor separation of components of the sample during the analysis. Overlapped peaks in UV-VIS or infrared absorbance, fluorescence, X-ray photoelectron (XPS), and nuclear magnetic resonance (NMR) spectroscopy represent overlapped spectral features. Hence, there is a real need to separate overlapped peaks in a reliable way to extract the information relevant to each component peak.

For most of the techniques except NMR, there is little in the way of fundamental theory on which to base models for nonlinear regression. Therefore, model building for peak-shaped data tends to be somewhat empirical. Nevertheless, various peak shapes can be employed reliably as models for such data, e.g., Gaussian and Lorentzian shapes. The Gaussian shape is derived from the normal curve of error, and the Lorentzian shape is more narrow at the top but wider at the very bottom of the peak (Figure 3.6). Models for a number of these peak shapes are listed in Table 3.8. The parameters are typically peak height, the half-width of the peak at half-height, and the position of the peak maximum on the x axis.

Models for overlapped peaks can be constructed by summing up two or more Gaussian, Lorentzian, or other peak-shape functions. For example, suppose we have chromatographic peaks that are approximated well by Gaussian peak shapes. Two overlapped peaks are represented by the sum of two Gaussian equations (Table 3.8).

Although for experimental data it might be necessary to add background terms or use a function that accounts for tailing [1], we illustrate the use of an overlapped peak model in the ideal case where none of these refinements to the model is necessary. The model for two overlapped Gaussian peaks (Table 3.8) was used to fit noisy simulated data for two overlapping

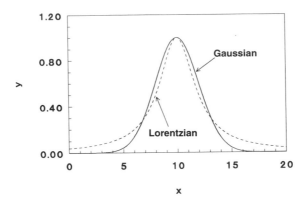

Figure 3.6 Gaussian (solid line) and Lorentzian (dashed line) peak shapes used for models of peak-shaped data.

peaks. The computed and measured data agree very well (Figure 3.7). Initial parameters were chosen as the best guesses possible by inspection of a graph of the raw data. Parameters and statistics are summarized in Table 3.9. We are able to recognize the characteristics of a good fit from this table. That is, SD $< e_y$ (0.005), and the deviation plot is random.

The use of nonlinear regression for peak resolution has the advantage that the component peaks can be reconstructed from the regression parameters. In this example, the computed estimates of height, width, and peak position are simply used with the equation for the shape of a single Gaussian peak (Table 3.8) to generate each component peak. In this way, accurate peak positions, heights, and areas (e.g., by integration) can be obtained for the components of the overlapped peaks.

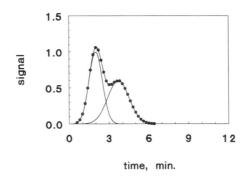

Figure 3.7 Deconvolution of Gaussian chromatographic peaks by nonlinear regression analysis. In outer envelope, points are experimental and line is the best fit from the analysis. Underlying peaks are component peaks computed from the results of the analysis.

Table 3.8 Models for Peak-Shaped Data

Peak shape	Model equation	Parameters
Gaussian	$(G)\ y = h\exp\{-(x - x_0)^2/2W^2\}$	h, x_0, W
Lorentzian	$(L)\ y = hW^2/[(x - x_0)^2 + W^2]$	h, x_0, W
Two overlapped Gaussian peaks	$y = b_1\exp\{-(x - b_2)^2/2b_3^2\}$ $+ b_4\exp\{-(x - b_5)^2/2b_6^2\}$	$b_1 = h_1,\ b_2 = x_{0,1},\ b_3 = W_1,$ $b_4 = h_2,\ b_5 = x_{0,2},\ b_6 = W_2$
Combined Gaussian and Lorentzian shapes	$y = h\{fG + (1 - f)L\}$	$h, x_0, W; f =$ fraction Gaussian
Gaussian model for n peaks with baseline	$y = mx + y_C + \sum_{i=1}^{n} h_i\exp\{-(x - x_{0,i})^2/2W_i^2\}$	$h_i, x_{0,i}, W_i$
n Gaussian/Lorentzian peaks with baseline	$y = mx + y_0 + \sum_{i=1}^{n} h_i\{f_i G_i + (1 - f_i)L_i\}$	$h_i, x_{0,i}, W_i, f =$ fraction Gaussian
Chromatographic peak with tailing (R vs. t data)	$R = \dfrac{1}{\tau}\int_{-\infty}^{t} dt'\, h\exp[(t - t_0)^2/2W^2] \times$ $\exp[(-(t - t' + t_C)/\tau]$	h, t_0, W, τ (see Chapter 14)
Chromatographic peak with tailing (R vs. t data)	$R = \dfrac{A}{\exp[-a(t - c)] + \exp[-b(t - c)]}$	A, c, a, b (see Chapter 14)

Symbols: h = peak height. x_0 = peak maximum location on x axis, W = peak width at half-height, t_0 = peak maximum on t axis, m = baseline slope, y_0 = baseline intercept.

Table 3.9 Results of Overlapped Gaussian Peak Model Fit onto Data Representing
Two Overlapping Peaks

Parameter/statistic	Initial value	True value[a]	Final value	Error
b_1	1	1.00	0.998	0.2%
b_2	2	2.00	1.998	0.1%
b_3	0.51	0.500	0.4996	0.08%
b_4	0.56	0.600	0.5987	0.2%
b_5	3.7	3.7	3.699	0.02%
b_6	0.9	0.80	0.804	0.05%
SD			0.0033	
Deviation plot			Random	

[a] Data generated by using the equation for two overlapped Gaussian peaks (Table 3.8)
with absolute normally distributed noise at 0.5% of the maximum y.

A bit of preliminary testing is often needed to find the best model for
peaks from a given experiment. This is best done with data representing
only single peaks. If we find the best model for single peaks, we can usually
apply it to the overlapped peak situation.

The literature provides considerable guidance in choosing models for
certain types of data. As mentioned earlier, successful fitting of chromatographic data has been done by using a Gaussian peak model convoluted
with a decreasing exponential in time after the peak to account for peak
tailing [1] (Table 3.8). The last entry in Table 3.8 shows another function
that accounts for tailing of chromatographic peaks (Chapter 14).

For NMR spectroscopy, the choice of a Gaussian or a Lorentzian peak
shape may depend on theoretical considerations (Chapter 8). Most peaks
in Fourier transform infrared spectroscopy can be reliably fit with Gaussian
or Gaussian–Lorentzian models (see Chapter 7).

A useful general model for peak shaped data involves a linear combination, of Lorentzian (L) and Gaussian (G) peak shapes (Table 3.8). A
parameter f is included as the fraction of Gaussian character. Thus, four
parameters per peak are usually used in the regression analysis. Product
functions of G and L have also been used [1]. A Gaussian–Lorentzian
model for n peaks is given in Table 3.8, with $mx + y_0$ representing the
slope (m) and intercept (y_0) of a linear background.

B.3. Models That Are Not Closed Form Equations

In many cases, models for experimental data cannot be expressed as
closed form equations; that is, no expression of the form $y = F(x)$ can be
found. The model may need to be solved for y by approximate numerical methods.

The lack of a closed form model need not be an obstacle to applying nonlinear regression, provided that a program is used that does not require closed form derivatives. A numerical model that computes $y_j(\text{calc})$ at the required values of x_j by any method can be used as the model subroutine of the regression program.

A relatively simple example is a model of the form $F(x, y) = 0$. An iterative approximation to find roots of such equations is the Newton–Raphson procedure [4]. This method can be used to solve polynomial equations such as

$$y^3 + a_1 y^2 + a_2 y + a_3 = 0. \tag{3.7}$$

The equation to be solved is of the form $F(y) = 0$. If an initial guess of the root is the value y_k, then the Newton–Raphson method holds that a better approximation, y_{k+1}, for the root is given by

$$y_{k+1} = y_k - F(y_k)/D(y_k) \tag{3.8}$$

where $F(y_k)$ is the value of the function at y_k and $D(y_k)$ is the first derivative of the function evaluated at y_k.

The method works well for polynomials if a reasonably good initial estimate of the root is provided. Equation (3.8) is then used repeatedly until convergence to a preset limit (**lim**) is reached. A typical convergence criterion is that the fractional change between successive values of y_k is smaller than the convergence limit, **lim**:

$$\text{abs}[(y_{k+1} - y_k)/y_{k+1}] \leq \textbf{lim} \tag{3.9}$$

where **lim** may be set at 0.0001 or smaller, as desired. Given a good initial estimate, convergence is rapid for well-conditioned applications and is often reached in a few cycles. Equations other than polynomials can also be solved by the Newton–Raphson approach.

A polynomial model describes the data in steady state voltammetry when the electrode reaction follows the stepwise pathway below:

$$O + e^- \rightleftharpoons R \tag{3.10}$$
$$2R \rightarrow R - R. \tag{3.11}$$

For this so-called EC^2 electrode reaction pathway (electron transfer followed by second-order chemical step), the product of the electron transfer at the electrode (eq. (3.10)) reacts with another of its own kind to form a dimer (eq. (3.11)). The steady state voltammogram is described by

$$y + [\exp(x - x_0)S] y^{2/3} - y_0 = 0 \tag{3.12}$$

where y is the current at potential x; x_0 is the standard potential for the O/R redox couple, $S = F/RT$ (F, T and R have their usual electrochemical meanings), and y_0 is the limiting current. In fitting this model to a set of

y, x data pairs, the Newton–Raphson method can be used to solve the model each time a computed value of y is needed by the nonlinear regression program.

This procedure will require the expressions

$$F(y_j, x_j) = y_j + [\exp(x_j - x_0)S]\, y_j^{2/3} - y_0 \quad \text{and} \quad (3.13)$$
$$D(y_j, x_j) = 1 + (2/3)\exp[(x_j - x_0)S]\, y_j^{-1/3} \quad\quad (3.14)$$

as well as eqs. (3.8) and (3.9). A procedure to compute the y_j(calc) values for this model in BASIC is given in the subroutine shown in Box 3.2.

The preceding is an example of a model that must be solved by numerical methods. A variety of numerical simulation methods may be used to provide models. For example, simulation models for electrochemical experiments have been successfully combined with nonlinear regression analysis [1]. In such cases, the simulation becomes the subroutine to compute the desired response.

The subroutine in Box 3.2 comes into play after the y and x data are read into the regression program. When the program requires values of

---

SUBROUTINE 3.2. BASIC CODE FOR NEWTON-RAPHSON METHOD

```
10      LIM = 0.0001: JJ%(I%) = 0

20      MM = (an expression to provide initial guess for each xj)

25      REM subroutine called by main program for each I% data point from 1 to
N%

30      F(I%) = MM + exp((X(I%)-B(0))*B(1))MM^(2/3) - B(2)

40      D(I%) = 1 + (2/3)*exp((X(I%)-B(0))*B(1))MM^(-1/3)

50      Y9(I%) = MM - F(I%)/D(I%)

60      REM test for convergence; iterations stopped if >50 cycles

70      IF ABS((Y9(I%) - MM)/Y9(I%))<=LIM THEN 120

80      MM = Y9(I%)

90      JJ%(I%) = JJ%(I%) + 1: IF JJ%(I%) > 50 PRINT "UPPER LIMIT OF
        CONVERGENCE EXCEEDED": GO TO 120

100     GOTO 30

120     YCALC(I%) = Y9(I%)
```

After line 120, return control to the main nonlinear regression program, to begin another estimate of YCALC(I%+1) or to continue the next cycle in the regression analysis if the end of the data set has been reached.

Box 3.2 BASIC code for Newton–Raphson method.

YCALC(I%), it must be directed to line 10 to obtain the Newton–Raphson solutions. The line by line description of the subroutine is as follows:

Line 10 sets a value for the tolerance limit LIM and sets a counter to zero.

Line 20 computes initial estimate of YCALC(I%). This can often be obtained as a multiple or fraction of the experimental *y* value.

Line 30 is the actual regression model (eq. (3.13)). It can be changed by the user.

Line 40 is the first derivative (eq. (3.14)) of the regression model, which requires modification if the model changes.

Line 50 computes Newton–Raphson estimate of the next best YCALC(I%).

Line 70 tests to see if the desired tolerance limit has been reached, and if so directs the program to compute final YCALC(I%) in line 120.

Line 80 defines the next "best guess" for YCALC(I%).

Line 90 increases the counter by 1, and tests to see if the maximum number of iterations has been exceeded. If so, the final estimate of YCALC(I%) will be made (line 120).

Line 100 directs program to begin the next Newton–Raphson cycle if the counter is smaller than the maximum number of cycles allowed.

Line 120 computes final YCALC(I%).

C. Finding the Best Models

C.1. Goodness of Fit Criteria

In many situations, the exact model for a set of data has not been established. The *goodness of fit criteria* for a series of nonlinear regression analyses to all of the possible models can be used to find the model that best fits the data. The usual goodness of fit criteria, deviation plots, standard deviations (SD), and error sums (S), are compared for this task.

Section 2.B.4 illustrated the use of SD, standard errors of parameters, and deviation plots for assessing goodness of fit. A SD smaller than the standard error in $y(e_y)$, assuming e_y is constant or absolute, along with a random deviation plot, can usually be taken as good evidence of an acceptable fit. If the number of parameters in every model examined is the same, the lowest SD consistent with an uncorrelated model and the best degree of randomness of the deviation plot can be used as criteria for choosing the best model. The use of summary statistics such as SD alone for such comparisons is risky, because summary statistics give little indication of systematic deviations of the data from the models. Deviation (residual) plots are often better indicators of goodness of fit than summary statistics because each data point is checked separately for its adherence to the regression model. The type of results needed to distinguish between models with the same number of parameters is illustrated in Table 3.10 for a hypothetical case with four models.

Models 1–3 in Table 3.10 all have SD $> e_y$ and non-random deviation plots. These are criteria for the rejection of these models. Only model 4 gives a random deviation plot and SD $< e_y$, indicating an acceptable model for the data.

If the models being compared have different numbers of regression parameters, summary statistics and deviation plots cannot be compared directly. A slightly different approach is needed to compare goodness of fit. Here, the individual regression analyses have different numbers of degrees of freedom, defined as the number of data points (n) analyzed minus the number of parameters (p). The difference in degrees of freedom must be accounted for when using a summary statistic to find the best of two models. A statistical test that can be used in such situations employs the extra sum of squares principle [1]. This involves calculating the F ratio:

$$F(p_2 - p_1, n - p_2) = \frac{(S_1 - S_2)/(p_2 - p_1)}{S_2/(n - p_2)} \tag{3.15}$$

where S_1 and S_2 are the residual error sums (S, eq. (2.10)) from regression analyses of the same data onto models 1 and 2, p_1 and p_2 now represent the numbers of parameters in each model, and the subscripts refer to the specific models.

To use eq. (3.15), model 2 must be a generalization of model 1. Regression analyses are done onto the two models, and the F statistic is calculated from eq. (3.15). The value of F obtained is then compared to the F value from tables at the desired confidence level, such as $F(p_2 - p_1, n - p_2)_{90\%}$. If the experimental $F(p_2 - p_1, n - p_2)$ is larger than $F(p_2 - p_1, n - p_2)_{90\%}$ but smaller than $F(p_2 - p_1, n - p_2)_{95\%}$ from the tables, then model 2 is the most probable model at a 90% confidence level. Different levels of

Table 3.10 Hypothetical Results from Analysis of a Single Data Set onto Four Different Models, Each with the Same Number of Parameters

Model	e_y	SD	Deviation plot
1	0.01	0.078	Nonrandom
2	0.01	0.033	Nonrandom
3	0.01	0.019	Nonrandom
4	0.01	0.0087	Random

confidence ($P\%$) can be employed until $F(p_2 - p_1, n - p_2)$ falls between two $F(p_2 - p_1, n - p_2)_{P\%}$, with the lowest of the two $P\%$s giving the confidence level with which model 2 is the most probable choice. If $F(p_2 - p_1, n - p_2) > F(p_2 - p_1, n - p_2)_{P\%}$, for all $P\% > 70$, then there is little confidence for concluding that model 2 fits the data better. In such cases, model 1 can be taken as the more appropriate model, because it explains the data with fewer parameters.

An example of the use of the extra sum of squares F test is illustrated in Table 3.11 for models that consist of a single exponential and the sum of two and three exponentials. The F test is done for the two- and three-exponential models. We see that the experimental F value is 2.79, which is in between the tabulated values of $F(2, 49)_{90\%}$ and $F(2, 49)_{80\%}$. This indicates that the three-exponential model can be accepted with at least an 80% confidence level. If the F value were smaller than $F(2, 49)_{70\%}$, this would reflect an insignificant difference between the error sums of the two models and model 2 would be the better choice.

Given the same degree of goodness of fit, the deviation plot will appear somewhat more random to the eye as the number of parameters is increased in related models. For fits with about the same SD for models of similar type, an increase in number of parameters tends to produce greater scatter in the deviation plots. This complicates direct comparisons between deviation plots for closely related models with differences between p_1 and $p_2 \geq 2$. However, with this knowledge in hand, an analysis based on deviation plots and the extra sum of squares F test can be used to distinguish between the models.

C.2. Dependent and Independent Variables Are Not Statistically Equivalent

In the preceding sections, we have identified the dependent variables in our models as the quantities measured in the experiment. The independent variable is time or the quantity that is controlled in the experiment. A few examples are listed in Table 3.12.

Table 3.11 Use of the Extra Sum of Squares F Test[a] to Distinguish between Summed Exponential Models

Model	No. parameters	S	$F(2, 49)$ exponential	$F(2, 49)$ table
1	2	1843		
2	4	69.01		@80% CL = 2.42
3	6	61.95	2.79	@90% CL = 3.19

[a] Analysis of 55 data points, see [1] for details.

Table 3.12   Examples of Variables in Typical Experiments

Experiment	Dependent variable, y	Independent variable, x
Spectrophotometry	Absorbance, intensity	Time, wavelength, frequency
Electron spectroscopy	Electron counts	Binding energy
Potentiometric titrations	pH or potential	Volume of titrant
Chromatography	Detector response	Time
Voltammetry	Current	Potential
Calorimetry	Heat absorbed or evolved	Time or temperature

It is important to realize that when nonlinear regression algorithms are used to minimize S in eq. (2.9), they are minimizing the errors in the dependent variable y with respect to the model. This can be seen by looking at the right-hand side of the equation:

$$S = \sum_{j=1}^{n} w_j [y_j(\text{meas}) - y_j(\text{calc})]^2 \qquad (2.9)$$

Only y appears there. This minimization ignores errors in x. For this reason, it is important to keep the identities of x and y consistent with their respective roles as independent and dependent variables. In practice, this means that the models should be written as $y = F(x, b_1, \ldots, b_k)$, and *not* as $x = F(y, b_1, \ldots, b_k)$. These equations are not statistically equivalent in terms of the minimized function S.

A simple example of the incorrectness of mixing up dependent and independent variables was discussed by Kateman, Smit, and Meites [5]. They showed what happens when the linear model

$$y = b_0 x + b_1 \qquad (3.16)$$

is converted to the algebraically equivalent form

$$x = b_0' y + b_1'. \qquad (3.17)$$

Thus, algebraically, we have

$$b_0 = 1/b_0' \quad \text{and} \quad b_1 = -b_1'/b_0'. \qquad (3.18)$$

However, when both eqs. (3.16) and (3.17) were subjected to linear regression onto the same data set, the equalities in eq. (3.18) were not followed. That is, if eqs. (3.16) and (3.17) were used, errors of $+0.4\%$ in b_0 and -2.1% in b_1 would be found. These errors resulting solely from the computation are too large to be acceptable. This example illustrates the principle that the identities of the independent and dependent variables, which are dictated by the nature of the experiment, must be kept intact during regression analyses.

C.3. The Number of Parameters in a Model

The number of degrees of freedom in a regression analysis was defined previously as the number of data points (n) analyzed minus the number of parameters (p). It is intuitive that the number of degrees of freedom should be significantly larger than 1. Although it might be acceptable to use a 5-parameter model to analyze a set of 6 data points, better accuracy and precision of the parameters would probably be obtained with 10 or 20 data points. On the other hand, there is usually an upper limit on the number of degrees of freedom needed to provide parameters of the required accuracy and precision. This can be tested for a given analysis by using data for a standard system with known parameters or theoretical data with a realistic amount of error added to it.

Some researchers in the physical sciences may hold the incorrect opinion that an increase in the number of parameters will always lead to a better fit of the model to the data. Solid arguments against this simplistic view have been presented in several review articles [1, 6]. Examples were presented to show that models of the wrong form with large numbers of parameters could not adequately fit data that was fit quite well by the correct model having fewer parameters.

An illustration of such an example is given in Figure 3.8, which graphically presents the results of a fit of two different polynomials onto a sigmoid curve generated with the equation

$$y = 1/\{1 + \exp[(1 - x)/0.0257]\} \qquad (3.19)$$

to which has been added 0.5% normally distributed absolute noise. This model is the same as that for steady state voltammetry in eq. (3.4). The

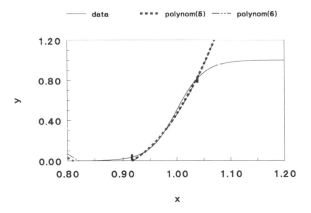

Figure 3.8 Plots of best fits of regression analysis of polynomial models with five and six parameters onto data representing a sigmoid curve described by $y = 1/\{1 + \exp[(1 - x)/0.0257]\}$, with 0.5% normally distributed absolute noise (solid line, 24 data points evenly spaced between the two solid vertical bars were analyzed). Polynom(5): $y = 14.2 - 25.8x + 2.60x^2 + 9.11x^3 + 0.354x^4$; Polynom(6): $y = 13.6 - 25.4x + 2.48x^2 + 9.11x^3 + 0.432x^4 + 0.233/x^2$.

Table 3.13   Results of Regression Analyses of Sigmoidal Data in Figure 3.8

Regression equation	No. parameters	10^2 SD^a	Deviation plot
$y = 1/\{1 + \exp[(1 - x)/0.0257]\}$	3	0.61	Random
$y = 14.2 - 25.8x + 2.60x^2$	3	3.9	Nonrandom
$y = 14.2 - 25.8x + 2.60x^2 + 9.11x^3$	4	2.3	Nonrandom
$y = 14.2 - 25.8x + 2.60x^2 + 9.11x^3 + 0.354x^4$	5	2.3	Nonrandom
$y = 13.6 - 25.4x + 2.48x^2 + 9.11x^3 + 0.432x^4 + 0.233/x^2$	6	2.3	Nonrandom

^{a} Standard deviation of the regression.

dashed lines in Figure 3.8 represent the best fits of five- and six-parameter polynomial equations onto the data. It is obvious that the fits of these polynomial equations are rather dismal. They simply do not fit the data. The model is incorrect.

The numerical results of these analyses (Table 3.13) show that the correct equation always gives the best standard deviation, regardless of the number of parameters in the *incorrect* polynomial equations. The four-parameter polynomial equation gives a better standard deviation than the three-parameter polynomial. However, the addition of further parameters does not improve the fit, as seen by the lack of improvement of the standard deviation for polynomials with four, five, and six parameters. Simply adding more parameters to the incorrect polynomial model does not provide a better fit!

The rather frivolous but often repeated statement that "with enough parameters you can fit an elephant" is exposed here as blatantly incorrect. If this statement *were* assumed to be correct, the results in Figure 3.8 and Table 3.13 would provide definitive, much needed proof that a sigmoid curve is not an elephant [6].

Of course, as implied by the discussion in Section C.1, inclusion of addition parameters in a model that already gives a relatively good fit has a tendency to lower the standard deviation a bit and introduce more scatter into the deviation plot. In such cases, the extra sum of squares test (eq. (3.15)) can be used to see if the apparent improvement in the fit is statistically significant.

References

1. J. F. Rusling, "Analysis of Chemical Data by Computer Modeling," *CRC Critical Reviews in Analytical Chemistry* **21** (1989), pp. 49–81.
2. J. F. Rusling, in S. J. Haswell (Ed.), *Practical Guide to Chemometrics*, pp. 151–179. New York: Marcel Dekker, 1992.

3. Y. Bard, *Nonlinear Parameter Estimation*, p. 316. New York: Academic Press, 1974.
4. K. J. Johnson, *Numerical Methods in Chemistry*, pp. 282–303. New York: Marcel Dekker, 1980.
5. G. Kateman, H. C. Smit, and L. Meites, "Weighting in the Interpretation of Data for Potentiometric Acid-Base Titrations by Non-Linear Regression, *Anal. Chem. Acta* **152** (1983), p. 61.
6. L. Meites, "Some New Techniques for the Analysis and Interpretation of Chemical Data," *Critical Reviews in Analytical Chemistry* **8** (1979), pp. 1–53.

Chapter 4

Correlations between Parameters and Other Convergence Problems

A. Correlations and How to Minimize Them

A.1. Recognizing Correlated Parameters

In Section 2.A.1 we discussed correlation between x and y in a linear regression analysis. This correlation is measured by using the product moment correlation coefficient, r. We found that x, y correlation in a linear regression analysis is a good thing. For a calibration plot, for example, we would usually like to achieve the largest possible x, y correlation, i.e., $|r| - 1$.

In Section 3.B.1 we mentioned briefly another kind of correlation; that is, correlation between two parameters in a regression analysis. This type of correlation is not desirable. If two parameters are fully correlated, unique and independent values cannot be found for each of them. A value found for one of these parameters will depend upon the value of the other one. Partial correlation between parameters can also occur and can pose serious problems if it is too large. We must avoid large correlations between parameters in nonlinear regression analysis.

How can we recognize correlation between parameters? In some cases, this is quite easy. For example, consider Table 4.1, which lists a series of models with correlated parameters. In entries 1–3 in this table, two parameters either multiply, divide, or are subtracted from one another. If parameters appear only in one of these relationships in the model, they are fully correlated and cannot be estimated independently by nonlinear regression analysis.

Table 4.1   Examples of Models with Correlated Parameters

Correlated model (1)	Correlated parameter	Corrected model (2)	Relation (1) ⟨=⟩ (2)
1. $y = b_0 b_1/(1 + b_2 x^2)$	b_0, b_1	$y = b_0/(1 + b_2 x^2)$	$b_0 = b_0 b_1$
2. $y = b_0(b_1 x^3 + \exp[-b_2 x])/b_3$	b_0, b_3	$y = b_0(b_1 x^3 + \exp[-b_2 x])$	$b_0 = b_0/b_3$
3. $y = b_0 \exp[(b_2 - b_1)x]$	b_1, b_2	$y = b_0 \exp[b_1 x]$	$b_1 = b_2 - b_1$
4. $y = F(b_0, b_1, \ldots, b_k)$, where $b_0 = f(b_1)$ or where model does not depend on b_0	b_0, b_1	$y = F(b_1, b_2, \ldots, b_k)$	$b_1 = F(b_0, b_1)$

In Table 4.1, $b_0 b_1$ (model 1), b_0/b_3 (model 2), and $b_2 - b_1$ (model 3) or $b_2 + b_1$ can be employed only as single parameters. The correct solution for this type of situation is to replace the correlated parameters in the model with a single parameter. This is shown in the "corrected model" column of Table 4.1; the relation between original and corrected model parameters is given in the final column.

We should not conclude that correlation exists anytime we find two parameters that are multiplied, divided, subtracted, or added in a model. For example, if model 1 in Table 4.1 were to have an additional term, so that

$$y = b_0 b_1/(1 + b_2 x^2) + b_1/x. \qquad (4.1)$$

We now have b_1 by itself in a term completely separate from $b_0 b_1$. This should allow the reliable estimations of each parameter, b_0, b_1, and b_2. Thus, full correlation can be assumed when the product (or quotient, sum, or difference) of parameters is the only place in which those parameters appear in the model. If another term exists in the model in which one of these parameters appears again, this usually removes the correlation.

The fourth model in Table 4.1 represents a general situation where one of the parameters exists in a mathematical relation with another parameter in the model. That is, if we can find an equation directly relating two of the parameters in a model, they are correlated.

In some cases, we may write a model for a set of data that has no information about a certain parameter in the model. An example is a current (y) vs. potential (x) curve for a completely irreversible electrode reaction controlled by electrode kinetics alone. The rising portion of this curve follows the expression

$$y = b_0 \exp[(x - E^0)/b_1] \qquad (4.2)$$

In this case, the standard potential E^0 cannot be used as an adjustable parameter because the data depend mainly on the kinetics of the electrode reaction. In other words, the data contain no information about the standard potential, which is a thermodynamic quantity. We would have to use an independently determined value of E^0 as a fixed parameter in this case.

Additionally, we can write an expression tying together b_0, E^0 and b_1, i.e.,

$$b_0' = b_0 \exp\{-E^0/b_1\}. \tag{4.2a}$$

Even from an algebraic point of view, eq. (4.2a) contains more parameters than any data set could provide. As mentioned above, a value found for one of these parameters will depend upon the value of the other one.

A similar situation arises if the same variability in the data is accounted for twice. For example, any simple exponential decay process which can be expressed as:

$$y = y_0 \exp\{-kt\} \tag{4.2b}$$

requires just two parameters, y_0 and k, to fully account for all variability. Attempts to tailor this model may lead the unwary into proposing a more "suitable" form, for example:

$$y = y_0 \exp\{-k(t - t_0)\}. \tag{4.2c}$$

We look at this apparently three-parameter model and ask ourselves: "What kind of variability do we account for with each parameter?" The presence of a completely correlated pair of parameters becomes obvious. For any data set consistent with the exponential decay model, we can choose several values of t_0, and while holding it constant, fit k and y_0. The model will fit equally well in each case, with the same summary statistics and identical deviation plots. Even the resultant k's will be the same. On the other hand, the values of y_0 will depend on the choice of t_0, since $y_0 = \exp\{kt_0\}$. If we allow t_0 to vary during a nonlinear regression routine, we may see matrix inversion errors in Marquardt–Levenberg approach. The convergence usually will take a little longer, but the final fit may be quite good. However, the final values of the parameters will depend strongly on the initial guess of the adjustable parameters. This is because the measured y contains no information about t_0. Thus, we need to be sure that the data contain information about all the parameters we are trying to estimate by nonlinear regression analysis.

In designing models for nonlinear regression, we may encounter situations where parameters will be at least partially correlated. The reason is that many functions we might want to include in the model based on our analysis of the physical nature of the experiment behave in similar ways in the range of independent variables. Look once again at the exponential model. Assume that we have determined the presence of a long-term linear drift in the instrument which we used to study a first-order decay. Appropriately, we modify our model to include this effect:

$$y = y_0 \exp\{-kt\} + mt + a. \tag{4.2d}$$

Here, the adjustable parameters are a, the value of the linear background at a time $t = 0$, m, the slope of the linear background, y_0, the deflection of the exponential from the background at $t = 0$, and k, the rate constant of the decay. If our largest value of t is significantly less than $4.6\ k^{-1}$, obtaining a reliable value for both k and m will be increasingly difficult. This problem will become very serious if our maximum t falls below $1/k$. The reason for this is quite clear from inspection of our data: the observed decay is practically a straight-line! In fact, now the exponential term is approximately

$$\exp\{-kt\} \approx 1 - kt. \tag{4.2e}$$

There is correlation between k and m. The algorithm we use for fitting the data cannot tell which variability to assign to what part of the model.

A good way to uncover more subtle correlations between parameters is by examining parameter correlation coefficients in a correlation matrix. A method of computing this matrix is described by Bates and Watts [1].

The diagonal elements of the symmetric correlation matrix are all unity. The off-diagonal elements can be interpreted like the correlation coefficient between y and x in a linear regression. If an off-diagonal element is close to 1, two parameters are highly correlated. Off-diagonal elements a_{ij} of the matrix, with i not equal to j, give an indication of the correlation between ith and the jth parameters. As a rule, correlation is usually not a serious problem if absolute values of off-diagonal elements are smaller than about 0.980. For example, if off-diagonal elements of 0.67 and 0.38 were found in a correlation matrix for a three parameter fit, we would conclude that there are essentially no correlations between parameters.

As an example of a correlation matrix, we will refer back to the sample regression analysis summarized in Table 3.7. Recall that data were fit onto a model of the form

$$y = b_1/(1 + \theta) + b_4 x + b_5 \tag{4.3}$$

where

$$\theta = \exp[(b_2 - E)b_3], \tag{4.4}$$

The correlation matrix elements obtained from this fit are shown (Table 4.2) only below the diagonal because it is a symmetric matrix. Elements above the diagonal are the mirror images of those below it. The bold numbers outside of the matrix are parameter labels. Thus, the correlation between parameters 2 and 3 is 0.045, that between 2 and 4 is 0.91 and so on. In this particular example, the only troublesome correlation is that between 1 and 4, at -0.991. This is a little bit above the arbitrary upper limit of safety we discussed previously. However, the other parameters show acceptable correlations, and the other goodness of fit criteria are

Table 4.2 Correlation Matrix for the Example in Table 3.7

1	2	3	4	5	
1					1
−0.94	1				2
−0.57	0.045	1			3
−0.991	0.91	0.065	1		4
−0.91	−0.80	0.059	−0.96	1	5

acceptable (see Section 3.B.1). Such a high correlation could be avoided by including more baseline data. This has not been done in the present case because the partial correlation creates no serious problems.

An easy way to find out if such "borderline" correlations cause significant errors is to compare calculations done with several sets of starting parameters. We have done this for the data used to generate Table 3.7. Results for a different starting point are listed in Table 4.3. The final parameter values of b_1 and b_4 have not changed very much in these two tables, and the errors in parameters b_1, b_2, and b_3 remain small. We can conclude that the partial correlation between parameters b_1 and b_4 is not highly significant in this analysis. The error in b_4 is large because of its small significance in determining y (see Section 3.B.1).

A.2. Orthogonalization of Parameters to Remove Correlation

If all else fails, serious *partial* correlation of parameters can be removed by orthogonalization of the parameter space. This orthogonalization method is general. In principle, it should be applicable to a wide variety

Table 4.3 Results of Model in Eq. (3.6) Fit onto Steady State Voltammetry Data with Background with Different Initial Parameters from Those in Table 3.7

Parameter/statistic	Initial value	True value[a]	Final value	Error
b_1	0.9	1.000	0.984	1.6%
b_2	39	38.92	39.66	2.0%
b_3	−0.2	−0.200	−0.1996	0.2%
b_4	−0.4	−0.200	−0.264	32%
b_5	0.005	0.005	0.0053	6%
SD			0.0033	
Deviation plot			Random	

[a] Data generated by using eq. (3.6) with absolute normally distributed noise at 0.5% of maximum y.

of models. However, it is not successful for models with fully correlated parameters.

Recall that in nonlinear regression the error sum S is minimized by systematic iterative variation of the parameters by an appropriate algorithm (Section 2.B.2). The approach to minimum S starting from a set of initial *guesses* for m parameters can be thought of as the journey of the initial point toward the global minimum of an error surface in an $(m + 1)$-dimensional coordinate system called *parameter space*. For a two-parameter model, we have a three-dimensional coordinate system. The z axis of this parameter space corresponds to S. The x and y axes correspond to the two parameters, respectively. The minimum S is called the *convergence point* (cf. Figure 2.4).

If correlation between two parameters is strong, convergence of the regression analysis or unique estimates of each parameter can be difficult to achieve. Parameter correlation is equivalent to having nonorthogonal axes in the parameter space (Figure 4.1). As an example of orthogonalization, we discuss parameter correlation that arises in models for obtaining rate constants of chemical reactions coupled to electrode reactions using a technique called *chronocoulometry*.

In chronocoulometry, a step of potential is applied to a working electrode, and the accumulated charge passing through the electrochemical cell

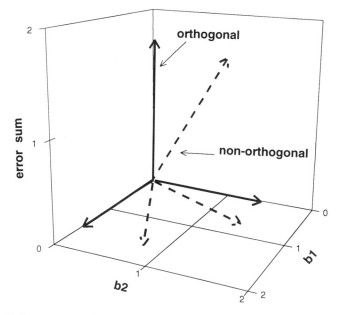

Figure 4.1 Representative orthogonal and nonorthogonal axes in three dimensions.

is recorded vs. time. Models describing charge response in chronocoulome-try can be expressed as explicit functions of time. Here, we summarize the application of the Gram–Schmidt orthogonalization to chronocoulometry of an electrode reaction occurring by an electron transfer–chemical reaction–electron transfer, or ECE, pathway [2]. We will return to the chemistry of ECE reactions in Chapter 12. At present we need to focus on only the form of the model.

The initial potential (E_i) of the working electrode is chosen so that no electrolysis occurs. At time $t = 0$, the potential is rapidly pulsed to a value where the first electron transfer is so fast that its rate does not influence the shape of the charge-time response. Under these conditions, the response of the experiment $Q(t)$ for the ECE model is of the form

$$Q(t) = b_2 + b_1[2 - (\pi/4b_0t)^{1/2} \operatorname{erf}(b_0t)^{1/2}] + b_3t. \qquad (4.5)$$

The $b_0, \ldots, b_3$ are regression parameters. The rate constant for the chemical reaction is $k = b_0$. This kinetic constant is generally the required parameter in this type of analysis. Identities of the other parameters are not relevant to our present discussion.

Gram–Schmidt orthogonalization was used to transform eq. (4.5) to the form

$$y_j = B_1h_1(t_j) + B_2h_2(t_j) + B_3h_3(t_j; B_0) \qquad (4.6)$$

where the $h_i(t_j)$ are orthonormal functions, the B_i are a new set of parameters that depend on the original $b_0, \ldots, b_3$, and $B_0 = k$. The details of the mathematical manipulations follow.

Orthogonalization defines a new orthogonal parameter space. Here, the most desired parameter is the rate constant B_0, which is determined by the regression analysis. The other original parameters b_i are computed from the B_i estimated in the regression analysis.

Gram–Schmidt orthogonalization of an ECE reaction model in single potential step chronocoulometry removed correlations between parameters and greatly improved the speed of convergence when using the Marquardt–Levenberg nonlinear regression algorithm [2]. The orthogonalized ECE model was used to estimate chemical rate constants for decomposition of unstable anion radicals produced during the reduction of aryl halides in organic solvents. The orthogonalized model converged three to four times faster than the nonorthogonal model.

A.3. Gram–Schmidt Orthogonalization—An Example

We now discuss the details of orthogonalization by using the example of the ECE model for chronocoulometry in eq. (4.5). Direct use of this equation led to slow convergence using steepest descent or Marquardt–

Levenberg algorithms, and indications of partial parameter correlation were obtained from correlation matrices.

A model can be considered a linear combination of functions that may contain nonlinear parameters. Gram–Schmidt orthogonalization [3] leads to an equivalent model from the explicit model for a single potential step chronocoulometric response for the ECE reaction. The orthogonalized model is a linear combination of orthogonal and normalized functions derived from the explicit ECE model.

As mentioned previously, the correlation between parameters can be thought of as lack of orthogonality between the parameter axes in the error surface plot. Such correlations can be expressed as the scalar products of appropriate pairs of unit vectors representing parameter axes in the parameter space. The scalar product of two axis vectors will be unity for full correlation and close to zero for uncorrelated parameters. As discussed already, a matrix containing correlation coefficients for all pairs of parameters can be constructed from the results of a regression analysis.

A representative correlation matrix for eq. (4.5) regressed onto a set of simulated data with 0.3% noise (Table 4.4) shows the correlation between parameters b_1 and b_2 at -0.981, suggesting partial correlation and significant nonorthogonality. As in the previous example, this method did not cause serious errors in final parameter values, but the convergence times using a program with a Marquardt algorithm were quite long. Matrix inversion errors were sometimes encountered. The analysis providing the matrix in Table 4.4 required 27 iterations for convergence.

The object of orthogonalization is to transform the nonorthogonal model in eq. (4.5) to the form of eq. (4.6). In the new model, the $h_k(t_j)$ are orthonormal functions, and the B_k are a new set of four adjustable parameters.

The starting basis functions are chosen from the overall form of eq. (4.5):

$$f_1(t) = 1 \qquad f_2(t) = t \qquad f_3(t) = 2 - (\pi/4kt)^{1/2} \, \mathrm{erf}(kt)^{1/2}.$$

constant   linear dependence   term containing erf

$$(4.7)$$

In developing the orthonormal basis function set, the next step is to accept one of the starting functions without a change. Hence, the first orthogonal

Table 4.4 Correlation Matrix for the Fit of Simulated Data onto the Nonorthogonal Model in Eq. (4.5)

b_0	b_1	b_2	b_3	
1				b_0
-0.874	1			b_1
0.813	-0.981	1		b_2
0.428	-0.810	0.834	1	b_3

function $g_1(t)$ was made identical to $f_1(t)$. Normalization of g_1 was then performed. The normalized function h_1 took the form

$$h_1(t) = g_1(t)/N_1 \tag{4.8}$$

where N_1 is a normalization factor such that

$$\sum_{j=1}^{n} [g_1(t_j)/(N_1)]^2 = 1. \tag{4.9}$$

Solving for N_1 gave the first function of the orthonormal basis set:

$$h_1(t) = g_1(t)/N_1 = n^{-1/2}. \tag{4.10}$$

According to Gram–Schmidt rules [3], the second function of the orthogonal basis set is a linear combination of the second function from the original model and the first orthonormal function. That is,

$$g_2(t) = f_2(t) - a_{21}h_1(t). \tag{4.11}$$

Here, the second term on the right-hand side of eq. (4.11) removes the component of $f_2(t)$ in the direction of $f_1(t)$. Imposing the orthogonality requirement of a zero scalar product of basis functions,

$$\sum_{j=1}^{n} g_2(t_j)h_1(t_j) = 0 \tag{4.12}$$

produced the a_{21} constant:

$$a_{21} = \sum_{j=1}^{n} f_2(t_j)h_1(t_j) = n^{-1/2}\langle t \rangle \tag{4.13}$$

where $\langle t \rangle$ is the average value of t. Substitution for a_{21} in eq. (4.11) gives

$$g_2(t) = t - \langle t \rangle. \tag{4.14}$$

Finally, normalization of g_2 yields

$$h_2(t) = (t - \langle t \rangle)\left(\sum_{j=1}^{n} t_j^2 - n\langle t \rangle^2\right)^{-1/2} \tag{4.15}$$

The last orthogonal function takes the form

$$g_3(t) = f_3(t) - a_{32}h_2(t) - a_{31}h_1(t) \tag{4.16}$$

The coefficients a_{32} and a_{31} are found from the requirement of orthogonality between g_3 and both h_2 and h_1. A complication arises here because the function f_3 contains the nonlinear parameter k, represented here as B_0. Thus, a_{32} and a_{31} are functions of B_0 and must be evaluated each time B_0 changes during the iterative optimization of the parameters by the regression analysis. Also, the normalization factor N_3 depends on B_0 and has to be reevaluated whenever the latter changed. To simplify the regression calculations the following expression was derived:

$$N_3^2 = \sum_{j=1}^{n} f_3^2(t_j) - a_{32}^2 - a_{31}^2. \qquad (4.17)$$

Finally,

$$h_3(t; B_0) = g_3(t; B_0)/N_3(B_0). \qquad (4.18)$$

Thus, the orthogonal model for the ECE reaction in chronocoulometry is given by eq. (4.6) with h_1, h_2, and h_3 from eqs. (4.10), (4.15), and (4.18). The parameters to be used for regression analyses are $B_0 = k$ (the same as b_0 in the nonorthogonal model of eq. (4.5), B_1, B_2, and B_3.

A detailed description of the applications of this orthogonal model can be found in the original literature [2]. Here, we briefly discuss its performance on the same set of data that gave rise to Table 4.4 using the nonorthogonal model. The orthogonal model converged in 4 iterations, compared to 27 for the nonorthogonal model. We see from Table 4.5 that the off-diagonal matrix elements are very small indeed. Orthogonalization of the model has effectively removed correlations between parameters and dramatically improves the rate of convergence.

A.4. Fourier Deconvolution

This method can improve the resolution of data before nonlinear regression analysis. It has been used successfully for pretreatment of spectroscopic data composed of a collection of severely overlapped peaks [4]. In practice, Fourier deconvolution increases the resolution of a spectrum by decreasing the width of the component peaks, while keeping their peak positions and fractional areas the same. The fractional area is the area of the component peak divided by the total area of all the component peaks. The user of the software specifies a resolution enhancement factor proportional to the desired amount of decrease in the half-width of the component peaks.

Fourier deconvolution increases the resolution of a spectrum, as shown

Table 4.5 Correlation Matrix for the Fit of Simulated Data onto the Orthogonal Model in Eq. (4.6)

B_0	B_1	B_2	B_3	
1				B_0
-9.6×10^{-11}	1			B_1
3.4×10^{-11}	2.2×10^{-11}	1		B_2
3.5×10^{-5}	9.0×10^{-13}	5.9×10^{-13}	1	B_3

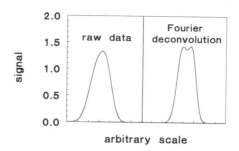

Figure 4.2 Illustration of a Fourier deconvolution for enhancing the resolution of raw data consisting of two severely overlapped peaks.

in the conceptual illustration in Figure 4.2. Programs for this procedure are available in the literature [5], in software provided by many infrared spectrometer manufacturers, and can be constructed by using Mathcad or Matlab (see the end of Chapter 2 for sources of these mathematics software packages) with their built-in fast Fourier transform capabilities.

Fourier deconvolution has been used extensively in the analysis of infrared spectra of proteins. In this application, the deconvolution aids in identifying the number of underlying peaks in the spectra. If the main interest of the researcher is to obtain peak positions and areas, nonlinear regression analysis of the Fourier deconvoluted spectrum should give identical results as the analysis of raw spectral data. We shall encounter further examples of the use of Fourier deconvolution in Chapter 7.

B. Avoiding Pitfalls in Convergence

B.1. Limits on Parameters

Many models used in a nonlinear regression analysis converge with no difficulty. In a few cases, however, minor problems may be encountered. In this section, we discuss some practical considerations concerning convergence, which should make the solution of these problems easier.

Our first concern involves the signs of parameters. Suppose we have a model in which all the parameters should be positive to make physical sense. For example, they might be concentrations of reagents, equilibrium constants, and rate constants. Obviously, these parameters cannot have negative values. Suppose we try to run a nonlinear regression analysis and find that one of the parameters, such as b_1, consistently goes negative. In extreme cases b_1 might stay negative, and the analysis might converge with a negative value of b_1. This would be a disconcerting result if, for example, b_1 represents a rate constant for a chemical reaction for which we have obtained products!

It should be realized in the preceding example that the program has found a minimum in the error sum in a physically unrealistic region of parameter space. Both of the solutions to this dilemma that follow strive to keep the value of the error sum away from physically meaningless regions.

The first solution (Table 4.6, method 1) involves simply using a logical statement in the program, in the part that computes y(calc), so that a parameter that becomes negative is forced to go more positive. This will increase the error sum for this cycle and hopefully redirect the search for the minimum in the correct direction. Another possible solution is to penalize the values of y(calc) to increase the error sum greatly if a parameter goes negative (Table 4.6, method 2). Both solutions can be tried in a given situation. In some cases, it may be necessary to force more than one parameter to remain positive. Appropriate logical statements similar to those in Table 4.6 can be employed. Similar approaches can be used to keep parameters negative.

B.2. Keeping Parameters Constant in Preliminary Analyses

Sometimes, a rather poor fit of the model or difficulty in convergence may be encountered if the model contains a large number of parameters and some of them are partially correlated. One example involves fitting an absorbance spectrum to a model containing five overlapped peaks. From Section 3.B.3, we know that we could use a model consisting of five Gaussian peaks with 15 parameters, consisting of peak widths, positions, and heights for each peak. However, a regression analysis that attempts to find all parameters at once will often fail to converge properly.

The solution to this problem is to obtain reliable estimates of the peak positions, such as by Fourier deconvolution or derivative spectra [4], then to fix the peak positions in a preliminary regression analysis. Once values of the 10 parameters describing widths and heights of the peaks are found in this preliminary analysis, they can be used as a starting point for the final nonlinear regression to find the best values of all the parameters. This technique will be discussed in more detail in Chapter 7.

Table 4.6 Program Statements Useful to Keep Parameters Positive

Method 1
IF $b_1 < 0$ then $b_1 = 1$ (or some other realistic positive number)
y(calc) = regression model

Method 2
y(calc) = regression model
IF $b_1 < 0$ then y(calc) = 2*y(calc)
Note: "IF" statements need to be placed in the routine that computes y(calc).

B.3. Error Surfaces*

In Section 2.B.1 we introduced the concept of the error surface. The global minimum of the error surface must be found for a successful regression analysis. Recall that the error surface has as its "vertical axis" the *error sum, S,*

$$S = \sum_{j=1}^{n} w_j [y_j(\text{meas}) - y_j(\text{calc})]^2 \qquad (4.19)$$

where the w_j are weighting factors, which depend on the distribution of errors in x and y. The other axes correspond to the parameters in the regression model. For a two-parameter model, we were able to visualize the error surface in a three-dimensional graph. Therefore, a nonlinear regression analysis was visualized as a journey to the minimum S on the error surface starting from a point corresponding to the initial parameter guesses (Figure 2.4).

We now consider the shapes of some real error surfaces, and how they can provide insight into convergence properties of a particular analysis problem. Consider the two-parameter linear model:

$$y = b_0 + b_1 x \qquad (4.20)$$

The error surface for this model is an elliptical paraboloid (Figure 4.3). This surface shows a sharp minimum. It features steep sides as we move away from the global minimum. It should provide no problems in convergence by any reliable minimization algorithm.

Nonlinear models can generate more complicated error surfaces than in Figure 4.3. In Figure 4.4, we show an error surface for an increasing exponential model:

$$y = b_0 \exp(b_1 x) \qquad (4.21)$$

This error surface does not have a sharp minimum but features a steep-sided, curved valley close to a rather shallow minimum. A region to the right and front of the minimum has a less steep slope.

A second model we shall consider describes the rate of an enzyme catalyzed reaction as a function of reactant concentration (cf. Section 6.B.3):

$$y = b_0 x / (b_1 + x) \qquad (4.22)$$

This error surface shows a somewhat broad valley that opens out into a relatively broader region near the minimum.

Error surfaces can be useful in understanding why a poor choice of initial values of parameters may result in a failure to converge or a matrix

---

* Section 4.B.3 was written by Artur Sucheta.

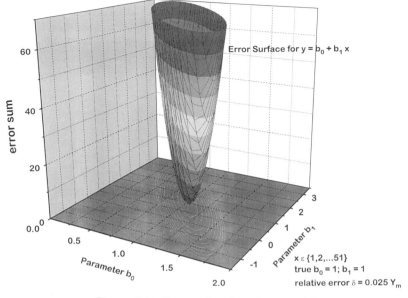

Figure 4.3   Error surface for a linear model.

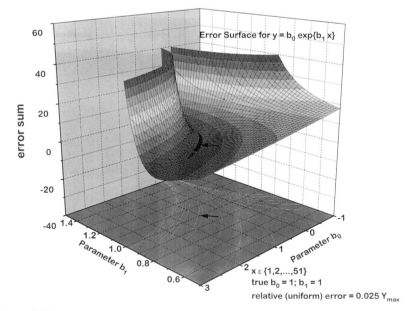

Figure 4.4   Error surface for an increasing exponential. Arrow points to global minimum.

inversion error. Figures 4.4 and 4.5 both show areas of the error surface that are relatively flat or parallel to the x, y plane. There are also regions with steep gradients. This may pose a problem for minimum-seeking algorithms in which one criterion for the direction of the subsequent iteration is the value of a gradient at the location of the current point on the error surface.

For example, programs based upon the rapidly converging Marquardt–Levenberg algorithm sometimes exhibit convergence problems when starting points are chosen in broad regions in the error surface close to the absolute minimum S. A typical program [6] compares two estimates of direction at each iteration. One is obtained from approximating the error surface as a parabola and determining the position of its minimum by matrix inversion [7]. The second estimate is obtained by evaluating the local gradient on the error surface. If these two directions differ by more than 90°, the discrepancy is deemed impossible to resolve and the program is terminated prematurely, returning a message of "matrix inversion error" to the user.

How can such a situation be resolved by a researcher eager to analyze a data set? Simply choosing a new initial set of parameters sometimes leads to convergence. The new initial set of parameters may give a better representation of the experimental data set or may be located in a more favorable region of the error surface. Thus, if the set of parameters defining the point P_i in Figure 4.5 gives convergence problems or premature program

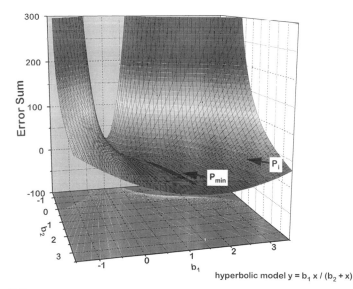

Figure 4.5 Error surface for a hyperbolic model. Arrow labeled P_{min} points to global minimum.

termination, an initial point on the other side of P_{min}, further back in the steeper valley, might possibly give better convergence properties and result in successful analysis of the data. At present, choosing the best starting point in such situations must be done by trial and error. Successful convergence depends on the shape of the error surface, and each error surface is a little bit different.

We have found that these types of convergence problems occur more frequently with programs based on the Marquardt–Levenberg algorithm than with other, slower algorithms. Perhaps because the Marquardt–Levenberg method is so rapid, it is less forgiving of unusual shapes of error surfaces.

In cases where the matrix inversion problem cannot be solved by a new set of initial parameters, a more conservative algorithm such as steepest descent or simplex (see Section 2.B.2) can usually be relied upon to find the global minimum, albeit with a larger number of iterations.

B.4. Propagation of Computing Errors

Because nonlinear regression is an iterative method, computational errors in a single cycle can be propagated into the final results. Although such propagation of errors is rarely a problem, it sometimes crops up when infinite series functions or built-in computer library functions are used to construct models [8]. The sine function on some computers sometimes causes propagation of errors, as can iterative or finite difference algorithms within the model. Error propagation can usually be eliminated or minimized by doing the computation in double precision.

In spite of their rather infrequent occurrence, we must make sure that computational errors do not bias the results of regression analyses. This can be done by analyzing sets of data simulated from the model that include various amounts of random noise (see Section 3.A), approximating the noise levels to be expected in the experiment. The parameters from such data sets are known, and the accuracy of the nonlinear regression method in determining them can be assessed before moving on to real experimental data [1, 8].

References

1. D. M. Bates and D. G. Watts, *Nonlinear Regression Analysis and Its Applications,* pp. 1–30. New York: Wiley, 1988.
2. A. Sucheta and J. F. Rusling, "Removal of Parameter Correlation in Nonlinear Regression. Lifetimes of Aryl Halide Anion Radicals from Potential-Step Chronocoulometry." *J. Phys. Chem.* **93** (1989), pp. 5796–5802.
3. G. Arfken, *Mathematical Methods for Physicists,* pp. 516–519. Orlando, FL: Academic, 1985.

4. J. F. Rusling and T. F. Kumosinski, "New Advances in Computer Modeling of Chemical and Biochemical Data," *Intell. Instruments and Computers* (1992), pp. 139–145.

5. J. K. Kauppinen, D. J. Moffatt, H. H. Mantsch, and D. G. Cameron, "Fourier Self-Deconvoluton: A Method for Resolving Intrinsically Overlapped Bands," *Appl. Spec.* **35** (1981), p. 271.

6. J. F. Rusling, "Analysis of Chemical Data by Computer Modeling," *CRC Critical Reviews in Analytical Chemistry* **21** (1989), pp. 49–81.

7. P. R. Bevington, *Data Reduction and Error Analysis for the Physical Sciences.* New York: McGraw-Hill, 1969.

8. S. S. Shukla and J. F. Rusling, "Analyzing Chemical Data with Computers: Errors and Pitfalls," *Anal. Chem.* **56** (1984), pp. 1347A–1368A.

PART II

Selected Applications

Chapter 5

Titrations

A. Introduction

This chapter deals with classical volumetric titrimetry [1]. In this collection of methods, the concentration of an analyte in a solution is determined by measuring the volume of a reagent of known concentration that is needed to react with that analyte. Traditionally, a reaction that goes to completion is required; for example,

$$A + R \rightarrow products. \tag{5.1}$$

The volume of the amount of reagent R (titrant) required to react with analyte A can be estimated by using an instrumental method to measure the concentration of A, R, or products (eq. (5.1)). A graph of these measurements vs. the volume of the titrant added to the solution is called a *titration curve*. Detection methods used involve measurements of pH, redox potential, conductivity, optical absorbance, and fluorescence.

In classical titrimetric analyses, the immediate goal is to find the *endpoint* volume of the reagent R as an estimate of the amount of R required to react completely with A. This allows an estimate of the amount of analyte A originally present in the sample, provided that the concentration of R is accurately known by a previous standardization.

In this chapter we shall see that some of the requirements of classical volumetric titrimetry are relaxed somewhat when nonlinear regression is used to analyze the data [2]. For example, the reagent R need not be accurately standardized, and the reaction need not go fully to completion at the endpoint. In fact, the endpoint of the titration need not even be determined. The concentration of A can be determined with high precision and accuracy directly from the titration curve. In the following sections, we provide a detailed example of nonlinear regression applied to the titration of a weak base with a strong acid. We then briefly discuss applications of the method to a variety of titrations employing different detection methods.

A.1. Titrations of Weak Acids and Bases

Models to analyze acid–base and other titration data employ the usual theory of equilibrium [1]. In general, an equation describing the measured response in terms of the initial concentration of analyte (c_a^0), concentration of titrant (c_b), initial volume of analyte (V_a^0), and volume of titrant added (V_b) is used as the model.

Weak Acids   Consider a potentiometric titrations of a weak monobasic acid with a strong base; that is,

$$HX + OH^- \rightleftharpoons H_2O + X^-$$

The exact equation [3] for the concentration of hydrogen ion during the titration is

$$[H+]^3 + \alpha[H+]^2 + \beta[H+] + K_a K_w = 0 \tag{5.2}$$

where

K_a = the dissociation constant for the weak acid,
K_w = the ion product $[H+][OH-]$,
c_a^0 = initial concentration of analyte,
c_b = concentration of titrant,
V_a^0 = initial volume of analyte,
V_b = volume of titrant added,
f = fraction of analyte titrated = $V_b c_b / V_a^0 c_a^0$,
$\alpha = c_a^0 f/(1 + rf) + K_a$,
$\beta = c_a^0 K_a (1 - f)/(1 + rf) + K_w$,
$r = c_a^0/c_b$.

The third-order polynomial in $[H+]$ is used along with the relation converting $[H+]$ to pH:

$$pH(calc) = -\log \gamma_{H+}[H+] \tag{5.3}$$

where γ_{H+} = the activity coefficient of hydrogen ion.

Equations (5.2) and (5.3) can be used in the model subroutine to compute a value of pH after each increment of titrant added. Solution of the third-order polynomial by an appropriate numerical method, such as Newton–Raphson (Section 3.B.3) is required.

Weak Bases   A second-order equation in $[H+]$ obtained by neglecting K_W (eq. (5.4)) was shown by Barry and Meites [4] to give excellent results for the titration of acetate ion with hydrochloric acid, using 3 M KCl to keep activity coefficients constant. This equation has the advantage that it can be solved by the quadratic method.

$$[H+]^2 + \alpha[H+] + \beta = 0 \tag{5.4}$$

where

K_a = the dissociation constant for the conjugate acid of the weak base,
c_b^0 = initial concentration of analyte,
c_a = concentration of titrant,
V_b^0 = initial volume of analyte,
V_a = volume of titrant added,
f = fraction of analyte titrated = $V_a c_a / V_b^0 c_b^0$,
$\alpha = K_a + (V_b^0 c_b^0 - V_a c_a)/(V_b^0 + V_a)$,
$\beta = V_a K_a c_a / (V_b^0 + V_a)$.

Ignoring the negative root, the quadratic solution to eq. (5.2) is

$$[H+] = \left[-\alpha + \sqrt{(\alpha^2 - 4\beta)} \right]/2. \tag{5.5}$$

A nonlinear regression program to analyze data from titrations of weak bases with strong acids requires eqs. (5.3) and (5.5) and definitions of α and β in the subroutine used to calculate pH as a function of V_a. A total of four *regression parameters* are possible:

$B(0) = K_a$
$B(1) = c_b^0$
$B(1) = c_a$
$B(3) = \gamma_{H+}$ an effective activity coefficient for hydrogen ion.

A subroutine for this calculation of pH as a function of V_a suitable for use in a nonlinear regression analysis is listed in Box 5.1.

When using this model, it is important to keep the ionic strength relatively large and constant, to keep γ_{H+} constant throughout the titration. This is best accomplished by using an inert 1:1 electrolyte such as NaCl at concentrations of 1–3 M.

Although this model contains four parameters, in few situations do we need to find all four of them in the analysis. We found that using this model with two or three parameters greatly improved convergence properties compared to four-parameter fits when using a Marquardt–Levenberg algorithm.

The most important parameter to be found in the titration is the analyte concentration c_b^0. The best way to find it by nonlinear regression is discussed later. We suppose that determination of c_a is also desirable, because then we can use an unstandardized reagent with a concentration known only to two significant figures. We can envision the following analytical situations.

Case 1   The identity of the weak base is known, and pK_a and γ_{H+} are known or can be found by a standard titration where both c_a and c_b^0 are known.

BASIC SUBROUTINE FOR COMPUTING pH (YCALC) IN TITRATION
OF WEAK BASE WITH A STRONG ACID

```
5 REM **YCALC subroutine ** FITS titration of weak base with strong
acid
20 ALPHA=(B(1)*VBASE-A(2,I%)*B(2))/(A(2,I%)+VBASE) + B(0)
30 BETA=(-A(2,I%)*B(0)*B(2))/(A(2,I%)+VBASE)
40 H=(-ALPHA+SQR(ALPHA^2-4*BETA))/2
50 YCALC = -LOG(B(3)*H)/LOG(10)
```

Documentation:

line 20 - computes α from $V_b{}^0$, V_a = A(2,I%), and values of parameters B(0), B(1),

and   B(2).

line 30 - computes β

line 40 - computes [H$^+$] from α and β using the quadratic formula

line 50 - computes pH from [H$^+$] and B(3).

Box 5.1 BASIC subroutine for computing pH (YCALC) in titration of a weak base with
a strong acid.

Finding pK_a and γ_{H^+} in this situation can be done experimentally by
titrating a solution of the analyte of known concentration with a standardized titrant. Reliable values of pK_a and γ_{H^+} can be found by fitting these
standard data to the model with c_a and c_b^0 fixed. These values of pK_a and
γ_{H^+} can then be used in subsequent titrations of real samples.

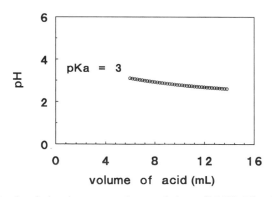

Figure 5.1 Simulated titration curve of a weak base (0.0100 M) with a strong acid
(0.0100 M) with γ_{H^+} = 0.90, pK_a = 3 for conjugate acid of the weak base, and an error of
±0.1% of a pH unit. Circles are simulated data, line is the best fit with model in Box 5.1.

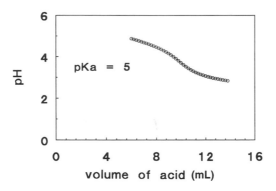

Figure 5.2 Simulated titration curve of a weak base (0.0100 M) with a strong acid (0.0100 M) with $\gamma_{H+} = 0.90$, $pK_a = 5$ for conjugate acid of the weak base, and an error of ±0.1% of a pH unit. Circles are simulated data, line is the best fit with model in Box 5.1.

We have followed the preceding scenario by using simulated data for titrations of 10 mL of 0.010 M weak base with 0.010 M strong acid. Data were simulated for various values of pK_a with the equations in Box 5.1, using $\gamma_{H+} = 0.90$. Absolute noise of 0.1% of a pH unit was added by using the subroutine in Box 3.1 in Chapter 3.

The shapes of titration curves computed with the above parameters for pK_a values of 3, 5, and 7 are given in Figures 5.1, 5.2, and 5.3, respectively. Data sets were computed with 40 points equally spaced on the V_a axis for extent of titration of 60% to 140%. This is the pK_a for the conjugate acid of the weak base. Thus, if X^- is the weak base, the values of pK_a pertain to the reaction:

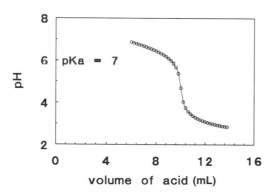

Figure 5.3 Simulated titration curve of a weak base (0.0100 M) with a strong acid (0.0100 M) with $\gamma_{H+} = 0.90$, $pK_a = 7$ for conjugate acid of the weak base, and an error of ±0.1% of a pH unit. Circles are simulated data, line is the best fit with model in Box 5.1.

$$HX \rightleftharpoons H^+ + X^-. \tag{5.6}$$

Since the basicity constant $pK_b = 14/pK_a$, the larger the pK_a, the stronger is the base X^-. Recall that the endpoint of a potentiometric titration is located at the inflection point of a sigmoid-shaped titration curve. For the data with $pK_a = 3$, the base is so weak that the shape of the titration curve is not sigmoid at all (Figure 5.1). This is expected because the pK_b is 11. The titration curve does not show any inflection point and remains in the lower pH region of the coordinate system.

When the pK_a is increased to 5, a rather poorly defined inflection point appears (Figure 5.2). At $pK_a = 7$, the base is strong enough for a reasonable inflection point (Figure 5.3), and the endpoint could be readily located by conventional derivative techniques [1].

Following the case 1 scenario, we determined pK_a and γ_{H^+} by nonlinear regression onto the model in Box 5.1, assuming that we know c_a and c_b^0 to an accuracy of $\pm 0.2\%$ in the experiment on standards. Then, in a separate set of analyses meant to emulate analysis of sample data, we used the values of pK_a and γ_{H^+} found in the first analysis in a new regression to obtain c_a and c_b^0. Results show (Table 5.1) that the errors in K_a and γ_{H^+} increase as pK_a increases; that is, as the base gets stronger.

The error in c_b^0 does not become significant, however, until we reach $pK_a \geq 7$. The error in c_b^0 appears to be most sensitive to the error in γ_{H^+} under these conditions. We shall see later that much better results are obtained by using three- or four-parameter fits.

Case 2   The identity of the weak base is unknown but γ_{H^+} can be found by a titration of a known base in which the ionic strength and electrolyte used is the same as in our analytical titration.

In this case, we assume that we can find γ_{H^+} by a one-parameter fit of the standard data, where pK_a for the standard known base, and c_a and c_b^0 are fixed at their known values. To find c_b^0 in the unknown samples, we now have the option of a three-parameter fit with γ_{H^+} fixed at its known value or a four-parameter fit. We have examined both situations for our test data. The results of the three-parameter fit were used as initial guesses in the four-parameter fits.

Table 5.1   Results of Initial Estimation of pK_a and γ_{H^+} Found with 0.2% Error in Concentrations, with Estimation of c_b^0 in Subsequent Two-Parameter Regression Analysis[a]

True pK_a	K_a found	γ_{H^+} found	c_b^0, found[b]	Error in c_b^0
3	1.001×10^{-3}	0.904	0.01001	+0.1%
5	1.002×10^{-5}	0.878	0.01007	+0.7%
7	1.010×10^{-7}	0.940	0.00848	−15%

[a] True values are $\gamma_{H^+} = 0.90$, $c_a = c_b^0 = 0.0100$ M.
[b] Initial guess of c_b^0 had +30% error.

The results of this analysis show that the approach works quite well for nearly all the data sets. Errors of <0.1% were found for all data with $pK_a \leq 7$, using either three- or four-parameter fits (Table 5.2). Only for a pK_a value of 9, that is, for the strongest base considered, was an error of 0.5% encountered. This latter data set required a two-parameter fit, because trouble with convergence of the Marquardt algorithm was encountered when three or four parameters were used.

Examination of Tables 5.1 and 5.2 suggests that an accurate value of γ_{H+} is important for the success of the regression analysis. We also find, at least with the Marquardt–Levenberg algorithm, that *the method works best for weaker bases.* Convergence problems are encountered when $pK_a \geq 7$. However, this does not pose a serious problem to the analyst. Conventional endpoint detection methods can be used in these cases, which are characterized by a clearly visible inflection point in the titration curve (Figure 5.3). It also appears that the approach taken in case 2 will give better results than in case 1. That is, even if the pK_a is known, the three- and four-parameter fits give better results than two-parameter fits.

Thus, we see that, as long as an accurate value of γ_{H+} can be obtained, excellent results can be expected for determination of the concentrations of very weak bases by titration with a strong acid. These titration-regression analyses can provide accurate values of analyte concentrations *without standardization of the titrant.* The concentration of the titrant need be known to only about two significant figures.

In practice, fitting the pH vs. V_b data by nonlinear regression can yield highly accurate and precise analyte concentrations and equilibrium constants. Typically, accuracy and precision within two parts per thousand can be achieved, even in titrations of weak bases such as acetate ($pK_b = 9.3$), which give poorly defined potentiometric endpoint breaks [4, 5].

Table 5.2 Results of Regression Analysis of Titration Data When γ_{H+} Is Known

True pK_a	No. parameters[a]	K_a found	γ_{H+} found	c_a, found	c_b^0, found	Error in c_b^0
3	3	1.000×10^{-3}	Fixed	0.01000	0.010001	+0.01%
	3[b]	1.016×10^{-3}	Fixed	0.01007	0.010001	+0.01%
	4	1.000×10^{-3}	0.900	0.01000	0.010001	+0.01%
5	3	9.995×10^{-6}	Fixed	0.009996	0.09996	−0.04%
	4	9.996×10^{-6}	0.899	0.009994	0.09993	−0.07%
7	3	1.000×10^{-7}	Fixed	0.009999	0.099998	−0.01%
	4	Matrix inversion error				
9	2[c]	1.29×10^{-9}	Fixed	Fixed	0.01005	+0.5%

[a] Number of parameters not fixed; true values are $\gamma_{H+} = 0.90$, $c_a = c_b^0 = 0.0100$ M. Initial guesses had ±30% error unless otherwise noted.
[b] Initial guess of c_b^0 had +90% error.
[c] Only two-parameter fits would converge for this set of data.

A.2. Survey of Selected Applications

A variety of potentiometric titration data that give rise to sigmoid-shaped titration curves can be analyzed by procedures similar to those already discussed. For example, Ingman and coworkers [6] showed that mixtures of weak acids with values of pK_a differing by only 0.12 units could be successfully determined with the aid of nonlinear regression analysis. Isbell and coworkers [7] applied the concept to precipitation titrations. Meites and Fanelli [8] showed that the reduced and oxidized forms of a redox couple could be determined from a single redox potentiometric titration. Meites [9] showed the possibility of detecting a 1% acidic impurity in solutions of an acid with a pK_a differing from that of the analyte by only 0.57 units. Weighting of the error sum reflecting errors in both measured pH and volume of titrant have been recommended [3]. Factors affecting the errors in such methods have been critically examined [10].

The *stability constant* is a fundamental characteristic of metal complexes. Titrations combined with nonlinear regression can be used to obtain stability constants of metal complexes. An excellent monograph on this subject is available [11].

Detection methods that directly measure the amount of a reactant, product, or titrant in a titration give data that can be described by the intersection of two straight lines. Methods falling into this category include spectrophotometric, conductimetric, and amperometric titrations. In the example shown (Figure 5.4), the species detected is the titrant. Only a small signal is observed prior to the endpoint. Beyond the endpoint, the increasing concentration of the titrant in the solution causes a linear increase in signal with volume added. Other possible arrangements of intersecting straight line shapes for these so-called segmented titration curves can be obtained,

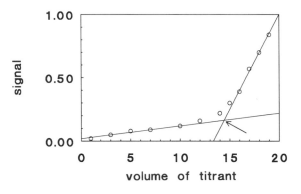

Figure 5.4 Simulated segmental titration curve characteristic of spectrophotometric, conductimetric, and amperometric titrations where the titrant is detected. The intersecting lines represent the model (Table 5.3) that can be used to fit such titration data and obtain the endpoint at their intersection, shown by the arrow.

Table 5.3   General Model for Linear Segmented Titration Curves

Assumptions: Titration data described by intersection of two straight lines; extreme
outliers removed, such as near the endpoint, x_0.
Regression equations:

$$y = m_1 x + b_1 \qquad \text{IF } x < x_0$$
$$y = m_2 x + b_2 \qquad \text{IF } x > x_0$$
$$x_0 = \frac{b_2 - b_1}{m_1 - m_2}$$

Regression parameters:	*Data:*
x_0 (endpoint) m_1 m_2 b_1 b_2	Response (y) vs. mL titrant (x)

depending on the species detected [1]. In such cases, models employing
two straight lines intersecting at the endpoint can be employed [12]. Such
a model is summarized in Table 5.3.

Thermometric titrations in thermostated vessels also give segmented titration curves when experimental conditions are controlled such that the
temperature changes are kept small. Shukla and Meites [13] combined
thermometric detection with acid–base titrations in micellar media and
analyzed the data by using an intersecting straight line model. In the presence of certain types of micelles in solution, some weak acids and weak
bases can be made stronger and are more easily titrated. By combining
this concept with nonlinear regression analysis of thermometric titrations,
the authors were able to determine simultaneously *ortho-* and *para*chloro-
phenol, which differ in pK_a by only 0.08 units in water.

References

1. D. A. Skoog, D. M. West, and F. J. Holler, *Fundamentals of Analytical Chemistry,* 5th
   ed. New York: Saunders, 1988.
2. L. Meites, "Some New Techniques for the Analysis and Interpretation of Chemical Data,"
   Critical Reviews in Analytical Chemistry **8** (1979), pp. 1–53.
3. G. Kateman, H. C. Smit, and L. Meites, "Weighting in the Interpretation of Data for
   Potentiometric Titrations by Non-linear Regression," *Anal. Chim. Acta* **152** (1983), p. 61.
4. D. M. Barry and L. Meites, "Titrimetric Applications of Multiparametric Curve-Fitting.
   Part I. Potentiometric Titrations of Weak Bases with Strong Acids at Extreme Dilution,"
   Anal. Chim. Acta **68** (1974), pp. 435–445.
5. D. M. Barry, L. Meites, and B. H. Campbell, "Titrametric Applications of Multi-Parametric
   Curve-Fitting: Part II. Potentiometric Titration with an Unstandardized Reagent," *Anal.
   Chim. Acta* **69** (1974), p. 143.
6. F. Ingman, A. Johansson, S. Johansson, and R. Karlson, "Titrations of Mixtures of Acids
   of Nearly Equal Strengths," *Anal. Chim. Acta* **64** (1973), p. 113.
7. A. F. Isbell, R. L. Pecsok, R. H. Davis, and J. H. Purnell, "Computer Analysis of Data
   from Potentiometric Titrations Using Ion-Selective Indicator Electrodes," *Anal. Chem.*
   45 (1973), pp. 2363–2369.

8. L. Meites and N. Fanelli, "Factors Affecting the Precision of a New Method for Determining the Reduced and Oxidized Forms of a Redox Couple by a Single Potentiometric Titration," *Anal. Chim. Acta* **194** (1987), p. 151.

9. L. Meites, "The Limit of Detection of a Weak Acid, in the Presence of Another, by Potentiometric Acid–Base Titrimetry and Deviation-Pattern Recognition," *Anal. Lett.* **15** (1982), pp. 507–517.

10. H. C. Smit, L. Meites, and G. Kateman, "Factors Affecting the Precision of Potentiometric Strong Acid–Base Titrations and Other Isovalent Ion Combination Titrations with Data Handling by Non-linear Regression," *Anal. Chim. Acta* **153** (1983), pp. 121–131.

11. A. E. Martell and R. J. Motekaitis, *The Determination and Use of Stability Constants.* New York: VCH Publishers, 1988.

12. J. G. McCullough and L. Meites, "Titrimetric Applications of Multiparametric Curve-Fitting: Location of End Points in Amperometric, Conductimetric, Spectrophotometric, and Similar Titrations," *Anal. Chim.* **47** (1975), pp. 1081–1084.

13. S. Shukla and L. Meites, "Thermometric and Other Titrations of Sparingly Soluble Compounds in Aqueous Micellar Media," *Anal. Chim. Acta* **174** (1985), pp. 225–235.

Chapter 6

Macromolecular Equilibria and Kinetics: Linked Thermodynamic Models

A. The Concept of Linked Functions

A.1. Introduction

Physical parameters that characterize macromolecular or colloidal systems often depend on the properties of different molecular species in equilibrium with one another [1]. Typical examples are encountered in experiments that use spectroscopic or electrochemical probes to characterize surfactant aggregates or protein solutions and in studies of the influence of salt on the solubility of proteins in solution. The general theory of *linked functions* or *thermodynamic linkage theory*, developed in the early 1960s by Wyman [2], provides a molecular and thermodynamic basis to analyze data from such systems.

The thermodynamic linkage approach can be used to relate an observed property (P_{obs}) to the properties of the individual components of the system that are linked together in coupled equilibria. For example, if P_j is the property we wish to investigate, and f_j is the fraction of the jth component in the system, a three-component system is represented by

$$P_{obs} = f_0 P_0 + f_1 P_1 + f_2 P_2. \tag{6.1}$$

In general, for a system with $k + 1$ contributing components,

$$P_{obs} = \sum_{j=0}^{k} f_j P_j. \tag{6.2}$$

Equations such as (6.1) and (6.2) can be used as models for nonlinear regression after the fractions f_j are expressed in terms of the relevant equilibria and equilibrium constants of the system [1]. In this chapter, we set out the general theory needed for obtaining the fractions f_j and discuss a number of examples of its use.

A.2. The Overall Equilibrium Approach

We begin by considering an overall equilibrium of the type

$$M + nX \rightleftharpoons MX_n. \tag{6.3}$$

If we equate chemical activities with concentrations, the equilibrium in eq. (6.3) can be described with an overall concentration equilibrium constant:

$$K^n = [MX_n]/\{[M][X]^n\}. \tag{6.4}$$

Equation (6.3) is identical to the classic representation of the overall equilibrium of a metal ion (M) with a ligand X. However, the representation is really more general. It can also be used to represent a macromolecule in equilibrium with bound anions or cations or a surfactant micelle in equilibrium with a solute used as a probe.

What eqs. (6.3) and (6.4) really represent are a series of linked or coupled equilibria. Suppose that $n = 4$ in eq. (6.4). Then we have the individual equilibria:

$$M + X \quad \rightleftharpoons MX \qquad K_1 = [MX]/\{[M][X]\} \tag{6.5}$$
$$MX + X \rightleftharpoons MX_2 \qquad K_2 = [MX_2]/\{[MX][X]\} \tag{6.6}$$
$$MX_2 + X \rightleftharpoons MX_3 \qquad K_3 = [MX_3]/\{[MX_2][X]\} \tag{6.7}$$
$$MX_3 + X \rightleftharpoons MX_4 \qquad K_4 = [MX_4]/\{[MX_3][X]\} \tag{6.8}$$

and the overall equilibrium constants is given by

$$K^n = K_1 K_2 K_3 K_4. \tag{6.9}$$

If the driving force for formation of MX_4 (or in general MX_n) is large, the intermediate equilibria in converting M to MX_4 may often be neglected. In such cases, the overall equilibrium can be used to construct models for the nonlinear regression analysis. For the $n = 4$ case,

$$M + 4X \rightleftharpoons MX_4 \qquad K^4 = [MX_4]/\{[M][X]^4\}. \tag{6.10}$$

The overall equilibrium model is applicable for many of the systems described in this chapter. Therefore, we now derive the expressions for the relevant fractions (f_j) of components for the overall equilibrium in eq. (6.3). We begin by expressing the total concentration of M (C_M) in the system as

$$C_M = [M] + [MX_n] \tag{6.11}$$

and the total concentration of X (C_X) as

$$C_X = [X] + n[MX_n].$$ (6.12)

The models to be developed require the fractions of bound and free M and X. In general, the fraction of a component j will be given by

fraction of j = (conc. free or bound j)/(total concentration of j). (6.13)

The fraction of free X $(f_{X,0})$ is expressed as

$$f_{X,0} = \frac{[X]}{[X] + n[MX_n]}.$$ (6.14)

From eq. (6.4), we have

$$[MX_n] = K^n[M][X]^n.$$ (6.15)

Substituting eq. (6.15) into (6.14) and canceling $[X]$ from numerator and denominator gives

$$f_{X,0} = \frac{1}{1 + n\,K^n[M][X]^{n-1}}.$$ (6.16)

The fraction of bound X $(f_{X,1})$ can be obtained by realizing that the sum of the fractions of all the forms of X must equal 1. In this case,

$$f_{X,0} + f_{X,1} = 1$$ (6.17)

and

$$f_{X,1} = 1 - f_{X,0}.$$ (6.18)

Alternatively, we can begin with the expression

$$f_{X,1} = \frac{n[MX_n]}{[X] + n[MX_n]}$$ (6.19)

and use the substitution for $[MX_n]$ in eq. (6.15) to yield the required expression for $f_{X,1}$ in terms of K^n.

Analogous procedures to those just discussed are used to obtain the fraction of free M $(f_{M,0})$ and the fraction of M with X bound to it $(f_{M,1})$. A list of these expressions for the overall equilibrium in eq. (6.3) is given in Table 6.1.

A number of other situations can be envisioned in which the linkage approach can be used to develop models. Suppose that two types of ions bind to a single macromolecule M but that X must bind before Y binds. The relevant overall equilibria are given in eqs. (6.20) and (6.21):

$$M + nX \rightleftharpoons MX_n \qquad K_1^n = [MX_n]/\{[M][X]^n\}$$ (6.20)
$$MX_n + mY \rightleftharpoons MX_nY_m \qquad K_2^m = [MX_nY_m]/\{[MX_n][Y]^m\}.$$ (6.21)

Table 6.1 Fractions of Bound
and Free M and X Derived Using
the Overall Equilibrium Model
in Eq. (6.3)

Fraction	Expression
$f_{X,0}$	$\dfrac{1}{1 + nK^n[M][X]^{n-1}}$
$f_{X,1}$	$\dfrac{nK^n[M][X]^{n-1}}{1 + nK^n[M][X]^{n-1}}$
$f_{M,0}$	$\dfrac{1}{1 + K^n[X]^n}$
$f_{M,1}$	$\dfrac{K^n[X]^n}{1 + K^n[X]^n}$

In this case, the measured property of the macromolecular solution might
depend on the fractions of all the forms M, MX_n, and MX_nY_n. The procedures outlined in eqs. (6.11) to (6.18) again can be used to obtain the
relevant fractions of each species. Examples similar to this one will be
described in applications later in this chapter.

A.3. Successive or Cooperative Equilibria

The general alternative to the overall equilibrium approach is to consider
all the individual equilibria involved. In the case of the system of equilibria
described by eqs. (6.5) to (6.8) we would need to consider the fractions of all
of the intermediate species from M to MX_4, as well as each of the equilibrium constants K_1, K_2, K_3, and K_4.

In general, such individual equilibrium steps can be described by

$$M + iX \rightleftharpoons MX_i \qquad i = 0 \text{ to } q. \qquad (6.22)$$

The total concentration of X is [2]

$$C_X = [X] + [M] \sum_{i=0}^{q} i\, K_i\, [X]^i. \qquad (6.23)$$

The total concentration of M is

$$C_M = [M] \sum_{i=0}^{q} K_i\, [X]^i. \qquad (6.24)$$

Procedures identical to those in Section A.2 can now be used to obtain
the relevant fractions, except that the definitions of total concentrations

Table 6.2 Fractions of Bound and
Free M and X Derived Using the
Successive Equilibrium Model
in Eq. (6.22)

Fraction	Expression
$f_{X,0}$	$\dfrac{1}{[M] \sum\limits_{i=0}^{q} i\,K_i [X]^i}$
$f_{X,1}$	$\dfrac{[M] \sum\limits_{i=0}^{q} i\,K_i [X]^i}{1 + [M] \sum\limits_{i=0}^{q} i\,K_i [X]^i}$
$f_{M,0}$	$\dfrac{1}{\sum\limits_{i=0}^{q} K_i [X]^i}$
$f_{M,1}$	$\dfrac{\sum\limits_{i=0}^{q} K_i [X]^i - 1}{\sum\limits_{i=0}^{q} i\,K_i [X]^i}$

are those in eqs. (6.23) and (6.24). The fractions for this general case are given in Table 6.2. Examples of more complex equilibria have been discussed in detail by Wyman [2] and Kumosinski [1].

B. Applications of Thermodynamic Linkage

B.1. Diffusion Coefficients of Micelles by Using Electrochemical Probes

Micelles are spherical or rod-shaped aggregates of surfactant molecules in water that can be characterized by their diffusion coefficients (D) [3]. Electrochemical measurements of current or charge in an electrochemical cell under an applied voltage can be used to estimate diffusion coefficients of the species being electrolyzed. A variety of specific electrochemical methods can be utilized for this purpose [3].

If D is available for a spherical micelle, its average radius (r) can be obtained from the well-known Stokes–Einstein equation:

$$r = kT/6\pi\eta D \tag{6.25}$$

where k is Boltzmann's constant, η is the viscosity of the solution, and T is temperature in Kelvins.

For a reversibly electrolyzed probe with a large binding affinity for the micelle, the electrochemically measured diffusion coefficient will reflect the equilibrium in eq. (6.3), with the overall equilibrium constant in eq. (6.4). Values measured in such systems are called *apparent diffusion coefficients* (D') because they depend on the equilibrium of the probe (X) with the micelle (M) [3, 4]. That is, D' depends on the amount of free probe in the solution.

The Single-Micelle Model Micelles actually have a distribution of sizes in aqueous solutions [5]. For a solution with a single narrow size distribution of micelles, where the probe–micelle equilibrium is rapid with respect to the measurement time, D' is given by

$$D' = f_{X,0}D_0 + f_{X,1}D_1 \qquad (6.26)$$

where the fractions are those defined in Table 6.1. D_0 is the diffusion coefficient of the free probe, and D_1 is the diffusion coefficient of the probe bound to the micelle. The binding (on) rates and dissociation (off) rates (cf. eq. (6.3) for small solutes in micellar systems are on the time scale of a few milliseconds to microseconds. The electrochemical methods used to measure D' are in the 10 millisecond to seconds time scale [3]. Therefore, eq. (6.26) is expected to hold for the majority of electrochemical probe experiments in micellar solutions with a single size distribution.

Insertion of fractions $f_{X,0}$ and $f_{X,1}$ from Table 6.1 into eq. (6.26) provides a model appropriate for analysis of the data (Table 6.3). If the goal of the data analysis is to obtain the diffusion coefficient of the micelle (D_1), D' vs. C_X data obtained at constant surfactant concentration (i.e., constant C_M) is the best choice. This is because micelle size generally depends on the surfactant concentration, so if surfactant concentration, and consequently C_M, is varied, the value of D_1 is no longer constant but depends on C_M. Only an average value of D_1 over the range of C_M used can be obtained when C_M is the independent variable.

An example of regression analysis onto the single micelle model in Table 6.3 using D' vs. C_X data for ferrocene as the probe shows an excellent fit for $n = 3$ (Figure 6.1). In these analyses, a series of nonlinear regressions

---

Table 6.3 Model for Single Size Distribution of Micelles

Assumptions: $[X] \ll C_x$ (probe tightly bound); K^n is an apparent equilibrium constant
Regression equation:

$$D' = \frac{D_0}{1 + K^n C_M C_X^{n-1}} + \frac{D_1 K^n C_M C_X^{n-1}}{1 + K^n C_M C_X^{n-1}}$$

Regression parameters: *Data:*
   D_0 D_1 $(K C_M^{1/n})^n$ D' vs. C_X or D' vs. C_M
Special instructions: Run a series of nonlinear regressions with successively increasing fixed integer n values beginning with $n = 2$ until the best standard deviation of regression is achieved

are run with n fixed at an integer value, beginning at $n = 2$. The n value at which the standard deviation of the regression is a minimum is taken as the best fit. In this case, for $n = 3$ the deviation plot was random. Values of $n = 4$ or 5 gave larger standard deviations.

The characteristic decrease in D' as C_X increases is explained by the single micelle model in Table 6.3 for 0.15 M hexadecyltrimethylammonium bromide in 0.1 M tetraethylammonium bromide (Figure 6.1), a system that most probably features rod-shaped micelles. A micelle diffusion coefficient of 0.33×10^{-6} cm^{2}s^{-1} was obtained from this analysis, which was similar to the value of 0.38×10^{-6} cm^{2}s^{-1} found for rod-shaped micelles of the same surfactant in 0.1 M KBr solution [3, 6].

The Two-Micelle Model   In applications to a series of surfactant solutions of concentrations above 0.05 M, modifications to the single micelle model were required to account for polydispersity caused by the presence of two size distributions of micelles. These systems typically contain spherical and rod-shaped micelles in the same solution [6], with the rod-shaped micelle having the smaller diffusion coefficient.

The following equilibria involving micelle 1 (M_1) and micelle 2 (M_2) need to be considered in the two-micelle model,

$$M_1 + nX \rightleftharpoons M_1X_n \tag{6.27}$$

and

$$M_2 + mX \rightleftharpoons M_2X_m \tag{6.28}$$

with overall concentration equilibrium constants

$$K_1^n = [M_1X_n]/\{[M_1][X]^n\} \tag{6.29}$$

and

$$K_1^m = [M_2X_m]/\{[M_2][X]^m\}. \tag{6.30}$$

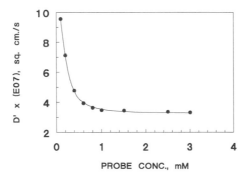

Figure 6.1 Influence of concentration of ferrocene (probe) on D' measured by cyclic voltammetry in 0.15 M hexadecyltrimethylammonium bromide/0.1 M tetraethylammonium bromide. Points are experimental data; line is the best fit to the data ($n = 3$) by the single micelle model (Table 6.3). (Adapted with permission from J. F. Rusling *et al.*, *Colloids and Surfaces*, 1990, **48**, 173–184, copyright by Elsevier.)

Now, three relevant species contribute to the apparent diffusion coefficient D'. The fractions of free probe ($f_{X,0}$), probe bound to M_1 ($f_{X,1}$), and probe bound to M_2 ($f_{X,2}$) must all be represented in the model. The expression for D' is

$$D' = f_{X,0}D_0 + f_{X,1}D_1 + f_{X,2}D_2. \tag{6.31}$$

The model for use in nonlinear regression was obtained by inserting the correct forms of the $f_{X,i}$ into eq. (6.31), and it is summarized in Table 6.4. When using this model, a series of fixed integer values of n and m must be tested until the n, m combination giving the minimum standard deviation of regression is found.

As with the single-micelle model, the best choice for experimental data when the micellar diffusion coefficients D_1 and D_2 are the desired parameters is D' vs. C_X. Regression analysis using the one- and two-micelle models is illustrated for a micellar solution 0.1 M in sodium dodecylsulfate (SDS) containing 0.1 M NaCl using methylviologen as the electroactive probe. As in any situation in which successively more complex models are suspected, the simple model should be applied first. Therefore, an initial series of regressions onto the single-micelle model gave the best standard deviation for $n = 3$. However, a graph of the regression line does not indicate a good fit to the data (Figure 6.2), and the deviation plot was distinctly nonrandom [6]. On the other hand, regressions of the same data onto the two-micelle model (Table 6.4) gave the best standard deviation for $n = 4$, $m = 8$. The graph of the regression line for the two-micelle model passed through all the data points (Figure 6.3) and gave a random deviation plot.

The two-micelle model contains five parameters and the one-micelle model has three parameters, so these two models will have different degrees of freedom for a given data set. It is necessary to use the extra sum of

Table 6.4 Model for the Diffusion of Two Different Sized Micelles in the Same Solution

---

Assumptions: $[X] \ll C_x$ (probe tightly bound); K_1^n and K_2^m are apparent equilibrium constants

Regression equation:

$$D' = \frac{D_0}{1 + K_1^n C_{M_1} C_X^{n-1} + K_2^m C_{M_2} C_X^{m-1}} + \frac{D_1 K_1^n C_{M_1} C_X^{n-1}}{1 + K_1^n C_{M_1} C_X^{n-1} + K_2^m C_{M_2} C_X^{m-1}}$$
$$+ \frac{D_1 K_2^m C_{M_2} C_X^{m-1}}{1 + K_1^n C_{M_1} C_X^{n-1} + K_2^m C_{M_2} C_X^{m-1}}$$

Regression parameters: *Data:*
  D_0 D_1 D_2 $(K_1^n C_{M_1}^{1/n})^n$ $(K_2^m C_{M_2}^{1/m})^m$ D' vs. C_x or D' vs. C_M

Special instructions: Run a series of nonlinear regressions with successively increasing fixed integer n and m values beginning with $n = 2$, $m = 2$ until the best standard deviation of regression is achieved which indicates the best values of n and m

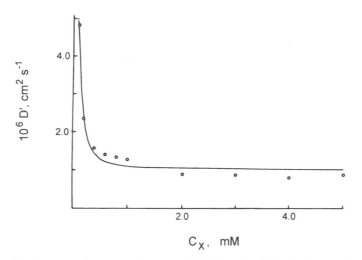

Figure 6.2 Influence of concentration of methyl viologen (probe) on D' measured by cyclic voltammetry in 0.1 M sodium dodecylsulfate/0.1 M NaCl. Points are experimental data; line is the best fit to the data ($n = 3$) by the single-micelle model (Table 6.3) (adapted with permission from [6], copyright by the American Chemical Society).

squares F test (Section 3.C.1) to examine the probability of best fit of the five-parameter vs. the three-parameter model. For the set of data in Figure 6.3, the error sum was 0.242×10^{-6} cm^{2}s^{-1} for the three-parameter model and 0.005×10^{-6} cm^{2}s^{-1} for the five-parameter model. Analysis using eq.

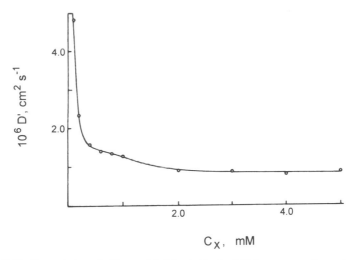

Figure 6.3 Same data as in Figure 6.2. Line is the best fit from regression onto the two-micelle model with $n = 4$, $m = 8$ (Table 6.4) (adapted with permission from [6], copyright by the American Chemical Society).

(3.15) with statistical F tables showed that the two-micelle model gave the best fit to the data with a probability of 99%.

Electrochemically determined diffusion data for number of micellar and microemulsion systems have been analyzed by the models in Tables 6.3 and 6.4. Results are in excellent agreement with diffusion coefficients of the same micelles estimated by alternative methods not requiring probes [3, 6]. Some of these data are summarized in Table 6.5. An advantage of the electrochemical probe method over many alternatives is that analysis for the presence of two micelles is relatively easy. The values of D_0 in these systems were similar to or a bit smaller than the D values of the probes in water. Slightly smaller values of D_0 than in water are expected because of the obstruction effect from the other micelles in solution [3, 4]. In some cases, D_0 may be influenced by association of the probe with submicellar aggregates [6].

B.2. Influence of Salt on the Solubility of Proteins

Salts have a large influence on the solubility of proteins. For example, low concentrations of salt in solution increase the solubility of the globular protein hemoglobin [7]. This is called *salting in*. As the salt concentration is increased, the solubility of hemoglobin decreases, and this is called *salting out*. Salts containing divalent ions are more efficient at these processes than salts with only monovalent ions.

Protein isolated from soybeans shows the opposite behavior to hemoglobin. Salting out occurs at low salt concentrations and is part of the basis for the production of tofu, a soybean curd product that is a staple food for much of the world's population. The solubility of soybean protein increases at higher concentrations of salt; that is, it salts *in*. Its solubility also depends upon pH.

The solubility of a protein can be thermodynamically linked to its capacity for binding salt. Solubility vs. [salt] profiles can be fit by nonlinear regression analysis by using models derived from thermodynamic linkage theory. The stoichiometry of salt binding and the binding equilibrium constants are obtained.

In models for the solubility of soybean protein, we assume that two classes of binding sites are responsible for the sequential salting-out and salting-in processes. The concept of linked functions is readily adaptable to building a model encompassing these processes. A similar situation involving two separate bound species was discussed briefly in Section A.3. In the present case, the equilibria are

$$M + nX \rightleftharpoons MX_n \qquad K_1^n = [MX_n]/\{[M][X]^n\} \qquad (6.32)$$
$$MX_n + mX \rightleftharpoons MX_nX_m \qquad K_2^m = [MX_nX_m]/\{[MX_n][X]^m\} \qquad (6.33)$$

ntra

efficients of Micelles from Nonlinear Regression Analysis of
a

Method[b]	No. of micelles	$10^6 D_1$ cm^2 s^{-1}	Lit.values[c] cm^2 s^{-1}	$10^6 D_2^d$ cm^2 s^{-1}
CV	2	1.41	1.40–1.45	0.84
CC	2	1.35	1.40–1.45	0.70
CV	2	1.45	1.40–1.45	0.99
CV	2	1.01	0.5–1.0	0.77
CV	2	0.73	0.5–0.83	0.38
CV	1	0.33^d	—	—
LSV^e	1	0.6	—	—

used to

lyzed di-
gram. As
iles were
est least-
at yielded

ein in the
s data for
at neutral
with the
curves are
perimental
errors are

chiometric
entative of
quilibrium
nding n or

y the same
, and NaI.
ies are also

; CTAB is cetyl- (or hexadecyl-) trimethylammonium
m bromide
) and ferrocene (Fc).
clic voltammetry; see [3, 4, 6] for details.
D values by alternative methods. Values listed assume

obtained by ultramicroelectrode linear sweep volta-
this fluid.

n, X is the free salt, n is the number of
ecies MX_n, and m is the number of moles
This model represents sequential binding
e n sites in eq. (6.32) are occupied before
sites on the protein (eq. (6.33).
depends on the solubilities of all of the

to S_0
ve to S_0.

measured solubility can be represented by

fferent fixed
is achieved

$$+ f_{1,M}S_1 + f_{2,M}S_2 \qquad (6.34)$$

where S_{app} = apparent protein solubility at a given total salt conce (C_X). The protein fractions are

$f_{M,0}$ = fraction of total protein as M
$f_{M,1}$ = fraction of total protein as MX_n
$f_{M,2}$ = fraction of total protein as MX_nX_m
$f_{M,0} + f_{M,1} + f_{M,2} = 1$
$C_M = [M] + [MX_n] + [MX_nX_m]$.

This overall equilibrium approach to the linkage problem was derive the model in Table 6.6.

Salt-dependent solubility profiles of soybean protein were ana rectly using a Gauss–Newton nonlinear regression analysis pro for the two-micelle model discussed previously, solubility pro analyzed by fixing the values of n and m and calculating the b squares fit. A series of n and m values were tested until those th the minimum error sum for the analysis were found.

Results are shown in Figure 6.4 for solubility of soybean prot presence of NaCl or NH_4Cl, NH_4Br and NaI. Figure 6.5 give NH_4NO_3 and $(NH_4)_2SO_4$ or Na_2SO_4. All plots show the best fits pH. The resulting regression lines are in excellent agreement solubility data. All deviation plots were random. The computed all within a relative standard deviation of $\pm 2\%$ from the exp data. The final parameter values with corresponding standard presented in Table 6.7.

It must be stressed that the K_j values are not the actual stoi equilibrium binding constants. They are average values repres only 1 mole salt bound to one protein site. The actual overall constants are calculated by raising the K_j value to the correspo m exponent.

As seen in Table 6.7, the salting-out constant, K_1, is essential within experimental error for NaCl, NH_4Cl, NH_4Br, NH_4NO Also, values for K_2 are similar for these salts. The n and m val the same; that is, $n = 1$ and $m = 2$ for all of these salts.

Table 6.6 Model for the Solubility of Proteins in the Presence of Salt

Assumptions: $C_M \ll [X]$, thus $C_X = [X]$
Regression equation:

$$S_{app} = \frac{S_0}{1 + K_1^n C_X^n} + \frac{S_1 K_1^n C_X^n}{1 + K_1^n C_X^n} + \frac{(S_2 - S_1)K_2^m C_X^m}{1 + K_2^m C_X^m}$$

Regression parameters: *Data:*
S_0 S_1 S_2 K_1^n K_2^m S_{app} vs. C_X
Special instructions: Run a series of nonlinear regressions with a series of di
integer n and m values until the smallest standard deviation of regression

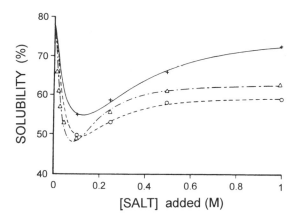

Figure 6.4 Influence of salt on the solubility of native soybean protein isolate. Points are experimental values and lines were computed from nonlinear regression analysis onto the model in Table 6.6: NaCl or NH_4Cl (O); NH_4Br (Δ); NaI (+) (adapted with permission from [1]).

The relatively low values of 1 and 2 for n and m, respectively, should not be interpreted literally as only a simple binding site, because multiple binding sites with nearly the same equilibrium constant would yield only a single binding isotherm. Hence, a value of n or m represents a class of protein binding sites rather than a single binding site linked to the solubility change of the protein.

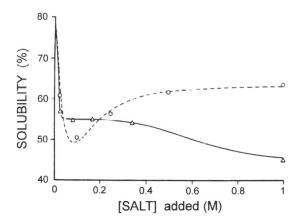

Figure 6.5 Influence of salt on the solubility of native soybean protein isolate. Points are experimental values and lines were computed from nonlinear regression analysis onto the model in Table 6.6: NH_4NO_3 (O); $(NH_4)_2SO_4$ or Na_2SO_4 (Δ) (adapted with permission from [1]).

Table 6.7 Parameters from the Nonlinear Regression Analysis of Solubility Data of Soybean Protein in Various Salt Solutions[a]

Salt	K_1	K_2	S_1, %	S_2, %	n	m
NaCl, NH$_4$Cl	41 ± 9	4.7 ± 0.4	31 ± 2	59 ± 4	1	2
NH$_4$Br	53 ± 16	6 ± 2	28 ± 10	62 ± 9	1	2
NH$_4$NO$_3$	51 ± 17	6 ± 2	29 ± 12	63 ± 13	1	2
NaI	50 ± 6	2.9 ± 0.3	43 ± 2	74 ± 4	1	2
Na$_2$SO$_4$, (NH$_4$)$_2$SO$_4$	100 ± 1	1.6 ± 0.1	55.1 ± 0.3	44 ± 1	4	4

[a] Data from [1].

The salting-out solubility, S_1, shows essentially no trend with respect to the type of anion used whereas a slight trend may exist for the salting-in solubility, S_2. Because soybean protein at neutral pH is thought to have a net negative charge, the preceding results can easily be interpreted in terms of an isoelectric binding model; that is, salt cations bind to negative sites on the protein surface, with an average equilibrium constant K_1, and produce a species of zero net charge with a corresponding solubility of S_1.

The salting-in of the protein at higher concentrations of salt can be thought of in terms of cations binding, with an average equilibrium constant K_2, to the unbound negative protein sites yielding a species, S_2, with a net positive charge. Alternatively, binding of both cations and anions of the salt to corresponding negative and positive protein sites can yield species with a net zero or negative charge. The values of S_1 and S_2 are slightly higher for NaI, 43 and 74, respectively, than the chloride, bromide, and nitrate values of approximately 30 and 60.

The salting-out solubilities of native soybean protein, S_1, are high in (NH$_4$)$_2$SO$_4$ and Na$_2$SO$_4$ in comparison with the other salts in Table 6.7. In addition, the K_1 value is significantly higher and the K_2 value is much smaller than for the other salts. The n and m values, both 4, were different from the $n = 1$ and $m = 2$ values for the other salts. Even though the shape of the solubility dependence on sulfate (Figure 6.5) is different from the rest of the salts, the model in Table 6.6 easily describes both types of solubility profiles. Applications of solubility models to other proteins have been discussed [1].

B.3. Enzyme Kinetics—Michaelis–Menten Model

The Michaelis–Menten equation is the standard model used to analyze kinetics of simple enzyme reactions [7]. The theory (shown in Box 6.1) assumes that the substrate (S) first binds to the enzyme (E). The enzyme–substrate complex then reacts to yield product P.

$$\begin{array}{c} k_{+1} \\ E + S \rightleftharpoons ES \\ k_{-1} \end{array} \qquad (6.35)$$

$$\begin{array}{c} k_{+2} \\ ES \rightleftharpoons E + P \\ k_{-2} \end{array} \qquad (6.36)$$

Box 6.1 Chemical basis of the Michaelis–Menten model.

The Michaelis–Menten constant K_M is the dissociation constant for eq. (6.35) and is given by

$$K_M = (k_{-1} + k_{+2})/k_{+1} = [S][E]/[ES]. \qquad (6.37)$$

The initial rate of an enzyme reaction is [7]

$$V_0 = k_{+2}[ES]. \qquad (6.38)$$

At high substrate concentrations, the enzyme becomes saturated, and $[ES] = C_E$, where C_E is the total enzyme concentration. The rate of the enzyme reaction is then at its maximum value V_{max}:

$$V_{max} = k_{+2}C_E. \qquad (6.39)$$

Using these definitions of V_{max} and K_M, the *Michaelis–Menten equation* takes the form

$$V_0 = \frac{V_{max}[S]}{K_M + [S]}. \qquad (6.40)$$

Equation (6.40) is usually linearized to eq. (6.41) for the estimation of V_{max} and K_M:

$$\frac{1}{V_0} = \frac{K_M}{V_{max}} \frac{1}{[S]} + \frac{1}{V_{max}}. \qquad (6.41)$$

A graph of $1/V_0$ vs. $1/[S]$ provides K_M/V_{max} as the slope and $1/V_{max}$ as the intercept on the $1/V_0$ axis. This is called a *Lineweaver–Burke plot* [7].

Data analysis problems created by linearization of equations were discussed in Section 2.A.3. In this case, we realize that the Michaelis–Menten model for enzyme kinetics can be formulated in terms of the expressions in Table 6.1. To make this clear, we express the scheme in Box 6.1 in terms of the symbols we have been using throughout this chapter. This is shown in Box 6.2. Because only bound enzyme leads to products, we can express the initial rate V_0 in the form:

$$V_0 = f_{M,1}V_{max}. \qquad (6.44)$$

$$M + X \underset{k_{-1}}{\overset{k_{+1}}{\rightleftharpoons}} MX \qquad (6.42)$$

$$MX \underset{k_{-2}}{\overset{k_{+2}}{\rightleftharpoons}} M + \text{Products} \qquad (6.43)$$

Box 6.2 Michaelis–Menten model using generalized symbols.

Expanding this expression using the definition of $f_{M,1}$ leads to the model in Table 6.8. The model is simply a different formulation of the Michaelis–Menten equation and in fact can be derived from eq. (6.40). The constant $K = 1/K_M$.

Farrell applied the model in Table 6.8 to the estimation of K_M and V_{max} for bovine $NADP^+$:isocitrate dehydrogenase in the presence of several monovalent cations [8]. Analysis of a particular set of data obtained from the author shows that the model provides a good description of the data (Figure 6.6). Nonlinear regression analysis gave $K_M = 2.26 \pm 0.26\ \mu M$ and $V_{max} = 163 \pm 3\ \mu\text{mol min}^{-1}\text{mg}^{-1}$ (relative units).

The traditional double reciprocal plot of the same data (Figure 6.7) shows considerably more scatter of data points around the regression line than the nonlinear regression results. Values obtained from the linear analysis were $K_M = 2.1 \pm 0.1\ \mu M$ and $V_{max} = 161 \pm 3\ \mu\text{mol min}^{-1}\text{mg}^{-1}$. These results are similar to those from nonlinear regression. We shall see in later sections of this chapter that the linear regression method has limited ability in the analysis of more complex enzyme kinetics, such as those involving inhibition.

B.4. The pH Dependence of Enzyme Activity

The influence of pH on the activity of an enzyme depends on the acid–base properties of the enzyme and the substrate. The shape of the pH vs.

Table 6.8 Model for Simple Michaelis–Menten Enzyme Kinetics

Assumptions: Michaelis–Menten enzyme kinetics (see [7]) $C_X \approx [X]$
Regression equation:

$$V_0 = \frac{KC_X V_{max}}{1 + KC_X}$$

Regression parameters: *Data:*
  $K(= 1/K_M)$ V_{max} V_0 vs. C_X
Special considerations:
  $K = 1/K_M$

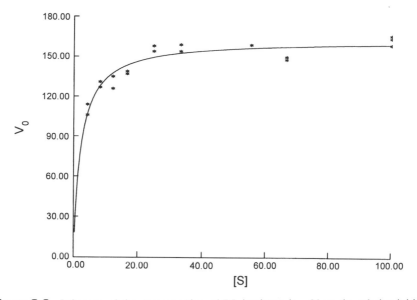

Figure 6.6 Influence of the concentration of DL-isocitrate in μM on the relative initial velocity of the isocitrate dehydrogenase catalyzed reaction in presence of 100 μM Mn^{2+}. Points are experimental data; line computed as best fit onto model in Table 6.8 (The authors thank Dr. H. M. Farrell for the original data.)

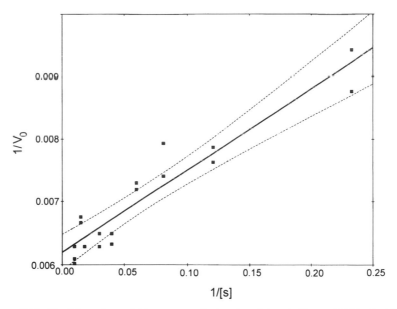

Figure 6.7 Double reciprocal Lineweaver–Burke plot of data in Figure 6.6 (Dashed lines show confidence intervals).

activity profile is often roughly bell shaped but may vary considerably depending on the system [7]. By working under conditions where the enzyme is saturated with substrate, we can employ the formalism that addition or removal or protons will change the activity of the enzyme.

Data on bovine $NADP^+$:isocitrate dehydrogenase will be used [8] to illustrate the linkage approach to analysis of an enzyme pH-activity profile. The roughly peak-shaped curve (Figure 6.8) suggests that an initial protonation activates the system, whereas a second protonation of this active form deactivates the system. The model can be expressed as

$$M + H^+ \rightleftharpoons MH^+ \qquad K_1 = [MH^+]/[M][H^+] \qquad (6.45)$$
$$MH^+ + H^+ \rightleftharpoons MH_2^{2+} \qquad K_2 = [MH_2^{2+}]/[MH^+][H^+]. \qquad (6.46)$$

This is a stepwise or cooperative model, and the K represents association constants. That is, one proton binds before the second one can bind. According to the linkage approach, the activity A_{obs} is given by [1]

$$A_{obs} = f_M A_1 + f_{MH} A_2 + f_{MH2}(A_3 - A_2). \qquad (6.47)$$

where A_1, A_2, and A_3 are the activities of M, MH^+, and MH_2^{2+}, respectively.

The full model is summarized in Table 6.9. This model was used to analyzed pH dependence data on the relative activity of bovine $NADP^+$:isocitrate dehydrogenase. The results are presented graphically in Figure 6.8. Parameter values from the analysis of these data follow:

$$\log K_1 = 8.6 \pm 0.2$$
$$\log K_2 = 7.0 \pm 0.4$$
$$A_1 = 0$$
$$A_2 = 68 \pm 14$$
$$A_3 - A_2 = -35 \pm 12$$

B.5. Influence of Multivalent Cations on Kinetics of a Cell Wall Enzyme

Plant cell walls contain many ionizable groups and may be regarded as immobilized polyelectrolytes. The ionic behavior of cell walls has been described by the well-known theories of Donnan and Gouy-Chapman.

The enzyme acid phosphatase, when bound to cell walls but not when solubilized in solution, is activated by increasing the ionic strength of the reaction mixture [1]. This apparent enzyme activation may be attributed to a decrease of the Donnan potential, which can inhibit the accessibility of negatively charged substrates to the cell walls. Acid phosphatase in plant cell walls may be important in hydrolyzing and solubilizing organic phosphate-containing macromolecules in the soil independent of microbial activity.

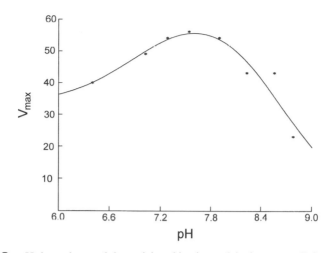

Figure 6.8 pH dependence of the activity of isocitrate dehydrogenase. Points are experimental data; straight line is best fit by linear regression. (The authors thank Dr. H. M. Farrell for the original data.)

The influence of multivalent cations on the activity of both bound and solubilized acid phosphatase associated with the primary cell walls of potato tubers and corn roots has been investigated. The results were analyzed in terms of binding equilibria using the thermodynamic linkage approach [1]. It was assumed that the observed change of acid phosphatase activity is due to a certain unique binding of cations to sites consisting of components of cell walls.

The analysis of the enzyme activity vs. cation concentration data assumed that molecules of acid phosphatase are uniformly distributed among many equivalent subdomains in the structure of the cell walls. Binding n metal ions to the subdomain, either directly to the enzyme, to its immediate environment, or to both, will cause a concentration-dependent change in the activity of the enzyme.

Table 6.9 Model for a Peak-Shaped Dependence of Enzyme Activity on pH

Assumptions: Two protonation equilibria as in eqs. (6.45) and (6.46)
Regression equations:

$$A_{obs} = \frac{A_1}{1 + K_1 C_H} + \frac{K_1 C_H A_2}{1 + K_1 C_H} + \frac{K_2 C_H (A_3 - A_2)}{1 + K_2 C_H}$$

$$C_H = 10^{-pH}$$

Regression parameters:	*Data:*
A_1 A_2 A_3 K_1 K_2	A_{obs} vs. pH

For the simplest case, consider the following binding equilibrium between the subdomain (M) containing the enzyme and metal ions (X):

$$M + nX \rightleftharpoons MX_n$$
$$K_a = [MX_n]/[M][X]^n \tag{6.48}$$

in which MX_n and M represent the domains with and without bound metal ions, respectively. K_a is the association equilibrium constant. These equations are the same as the simple overall equilibrium expressions in eqs. (6.3) and (6.4).

The dependence on metal ion concentration of the rate of enzymic hydrolysis of p-nitrophenyl phosphate (PNP-P) by cell wall acid phosphatase has been examined. By keeping the substrate concentration in large excess over the Michaelis constant K_M (i.e., $K_M = 1/K_a$), the observed apparent enzyme activity, A_{app}, for the hydrolysis of PNP-P in the presence of test cations, is expressed as

$$A_{app} = f_{M,0}A_1 + f_{M,1}A_2 \tag{6.49}$$

where A_1 is the activity of M and A_2 is the activity of MX_n. The fractions $f_{M,0}$ and $f_{M,1}$ are given in Table 6.1.

A_2 may be either greater or smaller than A_1. This can be expressed as

$$A_2 = QA_1. \tag{6.50}$$

Q is greater than 1 for enhancement of the activity and smaller than 1 for inhibition of the activity. The values of A_1 and C can be easily estimated from the activity vs. metal ion concentration data using the model in Table 6.10.

Plant cell walls contain relatively small amounts of protein, about 10% of the total dry weight of cell walls. Even assuming that all the protein in the cell walls is acid phosphatase, the maximum enzyme content would be very much smaller than the total concentration of metal ions (X) added

Table 6.10 Single Binding Site Model for Activity of Acid Phosphatase in the Presence of Metal Ions (X)

Assumptions: Domains of activity are uniformly distributed in cell wall.
$$C_X \gg C_M, \text{ so that } [X] = C_X$$
Regression equation:

$$A_{app} = \frac{A_1}{1 + K_a C_X^n} + \frac{QA_1K_aC_X^n}{1 + K_aC_X^n}$$

Regression parameters: *Data:*
   $A_1 \quad Q \quad K_a$ A_{app} vs. C_X

Special instructions: Run a series of nonlinear regressions with successively increasing fixed integer n values beginning with $n = 1$ until the best standard deviation of regression is achieved in the experiments. Hence the substitution $[X] = C_x$ is fully justified. The model based on eqs. (6.48), (6.49), and the preceding assumptions is given in Table 6.10.

The results of this analysis for cell wall acid phosphatase in the presence of several heavy metal ions are illustrated in Figure 6.9. The binding parameters are summarized in Table 6.11. In most cases, the best fits to the data were obtained by the use of the model in Table 6.10.

However, for Hg^{2+}-binding to potato cell wall, the model in Table 6.10 did not fit the data. The activity vs. $[Hg^{2+}]$ plot shown in Figure 6.9 indicated that the binding of Hg^{2+} at low concentrations enhances the activity. At higher concentrations of Hg^{2+}, inhibition becomes predominant. A model to account for this observation employed two independent binding sites for Hg^{2+} in the subdomains of the potato cell wall. Addition of n moles of X per mole of binding sites enhances the activity, whereas binding of

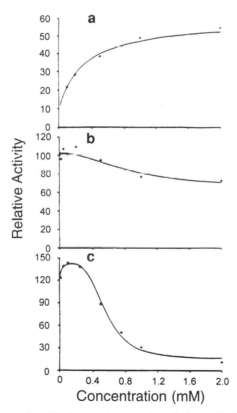

Figure 6.9 Influence of multivalent cations on the relative acid phosphatase activity of potato cell wall. (a) Mg^{2+}; (b) Al^{3+}; (c) Hg^{2+}. Points are experimental data and lines are best fits onto models discussed in text (adapted with permission from [1].)

Table 6.11   Metal Ion Binding Parameters for Cell Wall Acid Phosphatase[a]

Source	Ion	K_a, mM^{-n}	n	RSD, %[b]	Effect
Potato	Mg^{2+}	3.78	1	15	Stimulation
Potato	Al^{3+}	1.40	2	4	Inhibition
Potato[c]	Hg^{2+}	108.4	1	3	Stimulation
Potato[c]	(Second K)[c]	1.4×10^4	4		Inhibition
Corn	Hg^{2+}	4.82	1	2	Inhibition

[a] Best fit model in Table 6.10 unless otherwise noted.
[b] RSD refers to the average deviation between observed and calculated values of A.
[c] Two-site model in Table 6.11 for the effect of Hg on potato cell wall acid phosphatase.

an additional m moles of X inhibits the reactions. This binding model is represented as

$$M + nX \;\rightleftharpoons\; MX_n \qquad K_n \text{ (stimulation)}$$
$$M + mX \;\rightleftharpoons\; MX_m \qquad K_m \text{ (inhibition)}$$
$$X_nM + mX \rightleftharpoons X_nMX_m \qquad K_m \text{ (inhibition)}$$
$$MX_m + nX \rightleftharpoons X_mMX_n \qquad K_n \text{ (inhibition)}.$$

The observed activity, A, can then be expressed as

$$A_{\text{app}} = f_{M,0}A_1 + f_{M,1}QA_1 + f_{M,2}A_3 \tag{6.51}$$

where $f_{M,2}$ = fraction of M as X_mMX_n or X_nMX_m and $A_3 = Q'A_1$, with $Q' < 1$ to account for the inhibition at higher $[Hg^{2+}]$. This two-site model is described in Table 6.12.

Using the two-site model again necessitates searching for the integer values of n and m that give the lowest standard deviation of regression. The data for Hg^{2+} and potato cell wall acid phosphatase were best fit with $n = 1$ and $m = 4$. On the other hand, a sequential binding model that

Table 6.12   Two Binding Site Model for Activity of Acid Phosphatase in the Presence of Metal Ions (X)

Assumptions: Domains of activity are uniformly distributed in cell wall.
$$C_X \gg C_M, \text{ so that } [X] = C_X$$
Regression equation:
$$A_{\text{app}} = \frac{A_1 + QA_1K_aC_X^n + Q'A_1(K_aC_X^n + K_mK_nC_X^{m+n})}{1 + K_nC_X^n + K_mC_X^m + K_mK_nC_X^{m+n}}$$
Regression parameters: *Data:*
  $A_1 \quad Q \quad Q' \quad K_n \quad K_m$ A_{app} vs. C_X
Special instructions: Run a series of nonlinear regressions with successively increasing fixed integer n values and m values until the best standard deviation of regression is achieved

specified that the binding of the n sites occurred before the m sites did not provide a satisfactory fit to these data.

B.6. Enzyme Inhibition

B.6.a. Simple Competitive Inhibition

A competitive inhibitor can bind reversibly to active sites of an enzyme. In this way it competes with substrate for the active site and slows down the rate of the reaction [7]. A simple inhibition would be characterized, in the symbolism of Box 6.2, by the equation in Box 6.3, where I is the inhibitor, K_1^n is its binding constant to the enzyme M, and K_I is the dissociation constant for MI_n [7]. Inhibition can be studied by consideration of the combined effects of the schemes in Boxes 6.2 and 6.3 on the initial rate V_0. If the experiment is done at constant concentration of substrate and a series of concentrations of inhibitor, the appropriate expression is

$$V_0 = f_{M,0}V_{\max} + f_{I,1}V_1 \tag{6.54}$$

where $f_M,0$ (Table 6.1) denotes enzyme not bound to inhibitor, $V_{\max}$ was defined in Section B.3, $f_{I,1}$ is the fraction of enzyme bound to inhibitor, and V_1 is the apparent maximum velocity contributed by the MI state of the enzyme (this may be negative). The regression model is given in Table 6.13.

B.6.b. Bovine Isocitrate Dehydrogenase + Citrate

The velocity of the isocitrate dehydrogenase catalysis vs. concentration of the inhibitor Mn^{2+}-citrate shows a complex variation with C_I. Initial curvature upward is followed by gradual decline with a shoulder for C_I between 2 to 4 mM.

Preliminary analysis showed that the model was more complex than that in Table 6.13. Testing with a variety of models suggested that the data contained three distinct regions of response to Mn^{2+}-citrate. The set of chemical equations in Box 6.4 was found to lead to a model giving the best fit to the data [1].

Here $V_{\max}$ is the apparent $V_{\max}$ in the absence of concentration dependent binding of I, and V_1 through V_3 represent the velocity contributed by each state of the enzyme. The model implies three different binding sites for

---

$$M + nI \rightleftharpoons MI_n \tag{6.52}$$

$$K_1^n = [MI_n]/[M][I]^n = 1/K_I \tag{6.53}$$

Box 6.3 Basis of model for simple competitive inhibition.

Table 6.13   Model for Simple Inhibition of Enzyme Kinetics

Assumptions: Michaelis–Menten kinetics with inhibition as in Box 6.3.

$$C_I \approx [I]$$

Regression equation:

$$V_0 = \frac{V_{max}}{1 + K_1^n C_I^n} + \frac{K_1^n C_I^n V_1}{1 + K_1^n C_I^n}$$

Regression parameters: *Data:*
  K_1 V_{max} V_1 V_0 vs. C_I

Special instructions: Run a series of nonlinear regressions with successively increasing fixed integer n values beginning with $n = 1$ until the best standard deviation of regression is achieved

---

the inhibitor. Other models were tested but the fits were poor and not justified statistically. The binding of inhibitor to the enzyme M can be linked to the change in velocity using the linkage methods described earlier in this chapter. For data following the behavior in Figure 6.10, the model is summarized in Table 6.14.

The parameters obtained from analysis of the data in Figure 6.10 using the model in Table 6.14 shows that Mn^{2+}-citrate first stimulates activity characterized by $K_1 = 0.25$ mM ($n = 2$). Strongly inhibitory sites become bound at higher citrate concentrations with $K_2 = 3.83$ mM ($m = 2$), dramatically decreasing activity. As citrate concentration increases further, an apparent modulation of inhibition occurs, with $K_3 = 1.54$ mM ($q = 4$). The curve in Figure 6.10 is the composite of all these interactions. A complete discussion of the biological implications of these results has been given [1].

B.7. Interactions of Drugs with DNA

Binding sites and equilibrium constants for the binding of drug molecules to deoxyribonucleic acid (DNA) can be estimated by a technique called *footprinting analysis* [9, 10]. The experimental data that needs to be col-

---

V_{max}, K_1	V_1, K_2	V_2
$M + nI \rightleftharpoons$	$MI_n + mI \rightleftharpoons$	$MI_m I_n$
$\updownarrow K_3$	$\updownarrow K_3$	
$MI_n I_q + mI \rightleftharpoons$	$MI_m I_n I_q \ (V_3)$	

Box 6.4   Chemical equation model for inhibition of isocitrate dehydrogenase by Mn^{2+}-citrate.

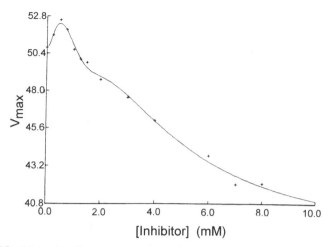

Figure 6.10 Main plot: V_0 at V_{max} conditions ($+$ moles/min/mg) vs. mM concentration of Mn^{2+}-citrate complex (M); the concentration of free-Mn^{2+} was fixed at 80 $+M$ and that of Mn^{2+}-isocitrate at 250 $+M$, variations of Mn^{2+}-citrate were then calculated. Data were fitted with model in Table 6.14 (adapted with permission from [1].)

lected is the rate of cleavage of the DNA at specific sites vs. the concentration of drug in the system.

The model for analysis of the data involves the combination of linked functions for all of the binding equilibria involving the drug and the DNA. The analysis is based on the fact that equilibrium binding of drugs to DNA may inhibit the DNA cleavage reactions at the specific site at which the drug is bound.

The footprinting experiment involves exposing the DNA to a drug bound by equilibrium processes, then cleaving the DNA backbone with the enzyme DNase I or some other cleavage agent [9]. If the equilibrium-binding drug

Table 6.14 Model for the Complex Enzyme Inhibition in Box 6.4

Assumptions: Inhibition mechanism as in Box 6.4.
$$C_I \approx [I]$$
Regression equation:
$$V_0 = \frac{V_{max}}{1 + K_1^n C_I^n} + V_1 \left[\frac{K_1^n C_I^n}{1 + K_1^n C_I^n} - \frac{K_2^m C_I^m}{(1 + K_2^m C_I^m)(1 + K_3^q C_I^q)} - \frac{K_2^m C_I^m K_3^q C_I^q}{(1 + K_2^m C_I^m)(1 + K_3^q C_I^q)} \right]$$
$$+ \frac{K_2^m C_I^m V_2}{(1 + K_2^m C_I^m)(1 + K_3^q C_I^q)} + \frac{K_2^m C_I^m K_3^q C_I^q V_3}{(1 + K_2^m C_I^m)(1 + K_3^q C_I^q)}$$

Regression parameters: *Data:*
V_{max} K_1 K_2 K_3 V_1 V_2 V_3 V_0 vs. C_I

Special instructions: Run a series of nonlinear regressions with successively increasing fixed integer n, m, and q values until the best standard deviation of regression is achieved

inhibits cleavage, the fragments of oligonucleotide products that terminate at the binding site will be present in smaller amounts in the product mixture obtained after the cleavage is quenched. Radiolabeled DNA is used, and analysis of the products is done by electrophoretic sequencing methods. Detection is by autoradiography. These experiments generate plots of auto-radiographic intensity of the oligonucleotide fragments vs. concentration of the drug [9, 10].

Although the model can be quite complex and may involve a large number of interacting equilibria, the binding can be explained by equations similar to the following:

$$M_j + X \rightleftharpoons M_jX \qquad K_j = [M_jX]/\{[M_j][X]\}. \qquad (6.55)$$

In the footprinting experiment, M_j represents the jth binding site on the DNA, X is the drug, and M_jX represents the drug bound to the jth site on DNA. The equilibria are expressed in concentrations of bound and unbound sites on DNA, so that

$[M_jX]$ = concentration of sites at which drug is bound,
$[X]$ = concentration of free drug,
$[M_j]$ = concentration of free sites without bound drug.

Using the same symbols and definitions as in Table 6.1, with $n = 1$,

$f_{M,0}$ = fraction of free sites,
$f_{M,1}$ = fraction of bound sites,
$f_{X,0}$ = fraction of free drug,
$f_{X,1}$ = fraction of bound drug.

Experimentally, the cleavage reaction must be terminated early so that each product results from only a single specific cleavage reaction of a DNA molecule. This yields product intensities in the analysis that are directly proportional to the probability of cleavage at the sites along the chain. If these conditions are fulfilled, the radiographic spot intensity is directly proportional to the rate of cleavage (R_i) at a given site, expressed as

$$R_i = k_i[A](1 - f_{M,1}) \qquad (6.56)$$

where k_i is the rate constant for cleavage at the ith site, and $[A]$ is the concentration of cleavage agent.

A given drug may bind at a site on the DNA encompassing several base pairs. For example, binding sites for actinomycin D (actD) span four base pairs [9]. For a position on the DNA or the nucleotide in a region where the drug binds, the drug blocks cleavage to an extent governed by its binding equilibrium constant at that site. In such cases, the cleavage rate (i.e., spot intensity) decreases as the concentration of drug increases [9]. On the other hand, for long DNA molecules with multiple sites, spot intensities for cleavage at sites *between* those that bind the drug increase with increasing concentration of the drug.

Structural changes in the DNA may be induced by drug binding. They may cause increases or decreases in cleavage rates that are not predicted by the simple picture of binding just discussed. Finally, sites having relatively small binding constants may exhibit intensity vs. $[X]$ plots showing cleavage rates enhanced at low drug concentrations and decreased at higher concentrations [9, 10].

Models for the DNA footprinting data are constructed based on eq. (6.56) combined with the definition of $f_{M,1}$ based on eq. (6.55). The preceding paragraph delineates the different types of qualitative behavior to expect in the data. In the discussion that follows, we give only an outline of how such models can be built. Full details of the model building and other complications of footprinting analysis are beyond the scope of this chapter, and the reader is referred to two excellent reviews articles on the subject [9, 10].

The essential feature of model building is to link eq. (6.56) to the correct expression for $(1 - f_{M,1}) = f_{M,0}$. This is then inserted into an expression for spot intensity (I_i), such as

$$I_i = k_i' [A](1 - f_{M,1}) \tag{6.57}$$

where k_i' is the product of k_i and the proportionally constants between I_i and R_i. Some examples of these expressions are given in Table 6.15.

Each system studied by quantitative footprinting is somewhat unique, and an appropriate model must be built starting from the factors summarized in Table 6.15. Additional factors may also have to be considered, such as the influence of unlabeled carrier DNA on the free drug concentration, as

Table 6.15 Components of Models for the Cleavage Rate Measured by Spot Intensity in DNA Footprinting

---

Independent binding sites:

$$I_{i,j} - k_i'[X]\left(\frac{1}{1 + K_j[X]}\right)$$

Adjacent binding sites k and j with mutual inhibition:

$$I_{i,j} = k_i'[X]\left(\frac{1}{(1 + K_j[X])(1 + K_k[X])}\right)$$

Adjacent binding sites k and j in which binding to one site excludes binding to the other site:

$$I_{i,j} = k_i'[X]\left(\frac{1}{1 + K_j[X] + K_k[X]}\right)$$

Concentration of free drug, where K_e is a regression parameter:

$$[X] = \left(\frac{1}{1 + K_e C_X}\right)$$

Regression parameters:	Data:
k_i' K_j K_k K_e	I_{ij} vs. C_X and site number (j)

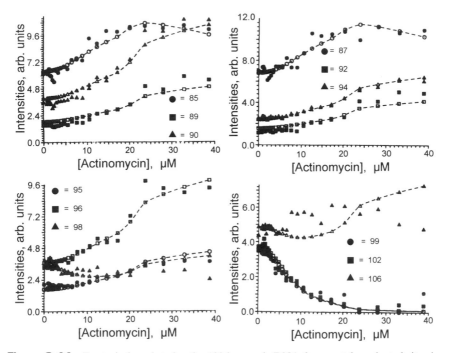

Figure 6.11   Footprinting plots for the 139 base pair DNA fragment for selected sites in the region 85–138 of the fragment. Solid symbols are experimental gel spot intensities. Fits computed according to the best model considering strong and weak binding sites are shown

well as weak binding sites [9, 10]. Nonlinear regression analysis is then done, with the error sum defined on the basis of the intensities of all the sites for all C_X. Thus,

$$S = \sum_{i,j} [I_{i,j}(\text{meas}) - [I_{i,j}(\text{calc})]^2, \tag{6.58}$$

I_{ij} is the dependent variable, and C_X and the site number j are the two independent variables. The data consists of sets of normalized intensities vs. C_X for each site, and all data are analyzed simultaneously. Simplex algorithms have been used to minimize S in this analysis.

The case of ActD with a 139 base pair DNA fragment, as well as other examples, have been discussed in detail [9, 10]. A later paper reported binding constants for both high and low affinity sites on the same 139 base pair DNA fragment [11]. The authors showed that both types of sites must be considered in the model, even if only the high-affinity binding constants are desired. Binding constants for 14 sites on the DNA were obtained. Examples of some of the graphical results presented in the work reveal

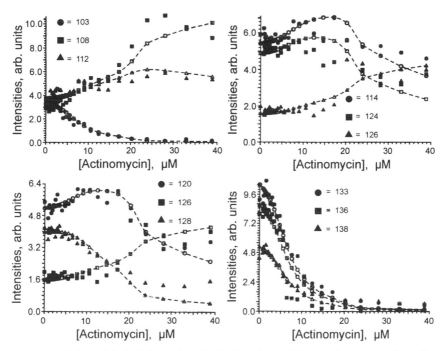

as open symbols and dashed lines Sites 89, 90, 95, 96, and 108 are enhancement sites. Sites 133, 136, and 138 show strong binding. (Adapted with permission from [11], copyright by the American Chemical Society.)

sites that exhibit strong binding and also those whose binding is enhanced as drug concentration is increased (Figure 6.11).

References

1. T. F. Kumosinski, "Thermodynamic Linkage and Nonlinear Regression Analysis: A Molecular Basis for Modeling Biomacromolecular Processes," *Advances in Food Science and Nutrition* **34** (1990), pp. 299–385.
2. J. Wyman, "Linked Functions and Reciprocal Effects in Hemoglobin," *Advances in Protein Chemistry* **19** (1964), pp. 223–286.
3. J. F. Rusling, "Electrochemistry in Micelles, Microemulsions, and Related Organized Media," in A. J. Bard (ed.), *Electroanalytical Chemistry,* vol. 19, pp. 1–88. New York: Marcel Dekker, 1994.
4. R. A. Mackay, "Electrochemistry in Association Colloids," *Colloids and Surfaces,* **82** (1994), pp. 1–23.
5. J. H. Fendler, *Membrane Mimetic Chemistry.* New York: Wiley, 1982.
6. J. F. Rusling, C.-N. Shi, and T. F. Kumosinski, "Diffusion of Micelle-Bound Molecules to Electrodes," *Anal. Chem.* **60** (1988), pp. 1260–1267.

7. A. L. Lehninger, *Biochemistry,* 2nd ed., p. 162. New York: Worth, 1976.

8. H. M. Farrell, "Purification and Properties of NADP⁺:Isocitrate Dehydrogenase from Lactating Bovine Mammary Gland," *Arch. Biochem. Biophys.* **204** (1980), pp. 551–559.

9. J. Goodisman and J. C. Dabrowiak, "Quantitative Aspects of DNASE I Footprinting," in *Advances in DNA Sequence Specific Agents,* vol. 1, pp. 25–49. JAI Press, Greenwich, C.T. 1992.

10. J. C. Dabrowiak, A. A. Stankus, and J. Goodisman, "Sequence Specificity of Drug–DNA Interactions," in C. L. Propst and T. J. Perun (Eds.), *Nucleic Acid Targeted Drug Design,* pp. 93–149. New York: Marcel Dekker, 1992.

11. J. Goodisman, R. Rehfuss, B. Ward, and J. C. Dabrowiak, "Site-Specific Binding Constants for Actinomycin D on DNA Determined from Footprinting Studies," *Biochemistry* **31** (1992), pp. 1046–1058.

Chapter 7

Secondary Structure of Proteins by Infrared Spectroscopy

A. Introduction

In this chapter we discuss the combination of nonlinear regression with auxiliary data analysis methods to extract secondary structural features of a protein from its infrared spectrum. Proteins are biopolymers made up of linear chains of amino acid molecules linked end to end in peptide bonds. They are also known as *polypeptides* [1]. The structures of an amino acid and a dipeptide with a peptide bond are as follows.

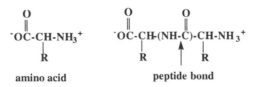

amino acid   peptide bond

The sequence of amino acids in the polypeptide chain is called the *primary structure.* The secondary structure of a protein is the way in which it is folded in the native state. Proteins fold in a complicated manner that is essential to their biological function. Proteins called *enzymes* are catalysts for life-supporting reactions whose activity and function depend intimately on their secondary structures. Other proteins serve as structural components of living organisms, and their function also relies on the secondary structure [1].

Secondary structures of proteins are characterized by helices, sheets, and extended regions within the polypeptide backbone. Other identifiable structural subunits include turns, loops, and disordered coils.

Secondary structural analysis of proteins can be done by X-ray crystallography, nuclear magnetic resonance spectroscopy (see Chapter 8), circular dichroism, and infrared spectroscopy. Among these methods, the technique of Fourier transform infrared (FT-IR) spectroscopy has emerged as a relatively straightforward technique for estimating secondary structural features of proteins in solution, provided the information is extracted from the data correctly [2–5]. The FT-IR method does not provide as detailed a structural picture of a protein as an X-ray crystal structure or an NMR analysis. However, it can be done rapidly on quite small amounts of material. The excellent signal to noise ratio, resolution, and accuracy of FT-IR has greatly facilitated this type of analysis.

A.1. Main Characteristics of Infrared Spectra of Proteins

The backbone of a polypeptide chain absorbs infrared radiation, which excites vibrational modes of the peptide bonds. Infrared spectroscopy measures the amount of light absorbed by these vibrations over a range of frequencies of the incident light. The positions and intensities of the infrared absorbance bands of these vibrational modes are sensitive to the protein's secondary structure. Band frequencies are characteristic of specific structural units, and their areas are proportional to the amount of the structural unit present in the overall structure.

Two vibrational modes are of prime importance for secondary structural analysis. The *amide I* mode is caused mainly by stretching of the carbonyl or CO bond [5]. Amide I vibrations appear in the general region around wave number 1660 cm^{-1}. The *amide II* mode is a combination of NH bend and CN stretch and appears in the region around 1540 cm^{-1}.

The positions of the amide I and amide II bands in the infrared spectrum are influenced by the exact structural unit in the polypeptide within which the bond resides. This leads to a number of severely overlapped bands within the amide I and amide II regions [5]. The individual bands in protein spectra can be modeled with the Gaussian peak shape, illustrated with its first and second derivatives in Figure 7.1. Torri and Tasumi [6–8] have calculated a theoretical FTIR amide I spectrum using the three-dimensional structure of lysozyme from X-ray crystallography and a Gaussian envelope of each peptide oscillator with a peak half-width at half-height of 3.0 cm^{-1}. The amide I and amide II bands of lysozyme after Fourier deconvolution are illustrated in Figure 7.2.

Kauppinen and coworkers [2] and Susi and Byler [3] considered the amide I band as a sum of severely overlapped Gaussian peaks, or Gaussian peaks with a fraction of Lorentzian character (see Section 3.B.2). Susi and Byler adapted a methodology using second derivative spectra, Fourier deconvolution algorithms (Section 4.A.4), and nonlinear regression analysis

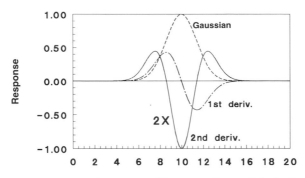

Figure 7.1 Shapes of a Gaussian peak and its first and second derivatives. The second derivative peak is plotted at twice the sensitivity of the other peaks.

for deconvoluting the amide I envelope into its individual component bands. Using the theoretical assignments for these bands [5, 9, 10], the fractions of various structural units such as α-helices, turns, extended structures, and so forth, in a protein's secondary structure can be estimated from the

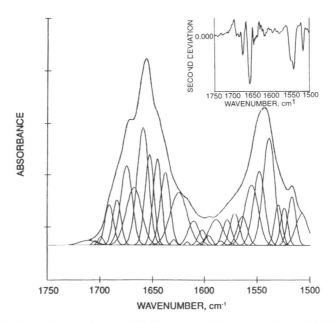

Figure 7.2 Fourier deconvolution of FT-IR spectrum of lysozyme in Figure 7.4. Overlapping lines on outer envelope show agreements of experimental and computed results. Individual component peaks underneath are the results of the regression analysis. Inset shows second derivative of original spectrum. (Reprinted with permission from [10], copyright by the American Chemical Society.)

fractional areas under the assigned component bands. Initially, the amide I band was considered only because experiments were done in D_2O, in which the amide II band does not appear. More recently, the amide I and amide II bands for spectra obtained in water and have been analyzed together [5, 9, 10]. As shown later, these analyses must be done in a way that is consistent with theoretical models [5–8] and statistical considerations.

A.2. Experimental Considerations

Fourier transform infrared spectroscopy employs an interferometer to collect an entire spectrum over the full range of energies simultaneously. This provides a more rapid collection of spectra than scanning through the required energy range, as is done by dispersive spectrophotometers. Thus, the FT-IR spectrometer collects many spectra in a short time, improving the signal to noise ratio by an amount proportional to the square root of the number of scans [11]. After the interferometric spectra are collected and averaged, the inverse Fourier transform is used to compute a vibrational spectrum, usually in the form of transmittance or absorbance vs. frequency in wavenumbers, from the interferogram. In modern spectrometers, this transformation is done automatically with software provided along with the FT-IR instrument.

FT-IR spectra can be obtained on many types of protein samples, such as solutions, gels, films, or colloidal dispersions. FT-IR spectra can be collected from microliters of protein solutions in the concentration range of 20–50 mg mL^{-1}. In the past, the infrared (IR) spectra of proteins were obtained in D_2O, but FT-IR spectrometers make it possible to obtain the spectra in H_2O. This offers considerable advantages, because both amide I and II bands appear in H_2O, providing a larger information content enabling a more complete analysis of the secondary structure.

Careful subtraction of the water spectrum from the spectrum of the protein solution is required when experiments are conducted in water. This subtraction should yield a horizontal line in the 1800–2000 cm^{-1} region, where only water and not protein absorbs IR radiation. The spectrometer must be purged thoroughly with dry nitrogen or argon to remove water vapor and carbon dioxide before collection of the spectrum.

The following experimental conditions were used in the examples cited in this chapter [9, 10]. Aqueous solutions were prepared as 4% protein in compatible buffers. All samples were introduced into a demountable cell with CaF_2 windows separated by a 12 μm Teflon spacer. Spectra were obtained using a Nicolet 740 FT-IR spectrometer equipped with the Nicolet 660 data system. Nitrogen purging of the sample chamber was used to reduce water vapor and carbon dioxide to a minimum. Each spectrum consisted of 4096 interferograms, coadded, phase-corrected, apodized with the Happ–Genzel function, and inverse Fourier transformed. Nominal instrumental resolution was 2 cm^{-1}, with one data point every 1 cm^{-1}. Water vapor absorption was routinely subtracted from all spectra.

A.3. The Model and Auxiliary Data Analysis Methods

Because of the large number of peaks under the amide I and amide II bands of proteins [5–8], auxiliary methods are required to aid in the resolution of the spectra. The nonlinear regression analysis must be assisted by relatively accurate initial choices of the parameters.

In our examples, second derivatives of the spectra were used to obtain initial peak positions on the frequency axis and to identify the number of peaks. Software to obtain derivative spectra are usually available in an FT-IR spectrometer's software package. A regression model consisting of a sum of Gaussian peaks, similar to that listed in Table 3.8, gave the best fit to protein and polypeptide spectra [9, 10].

Therefore, the first step of the analysis is to examine the second derivative of the spectrum and extract from it the approximate number of peaks in the amide I and II envelopes, along with initial guesses of their positions. To use the second derivative spectra, we should be familiar with the shape of the second derivative of a Gaussian peak. Shapes of a Gaussian peak and its first and second derivatives are given in Figure 7.1. The second derivative has a characteristic negative peak at the identical position on the x axis as the original Gaussian peak. Therefore, one negative peak should be in the second derivative spectrum for each component peak in the original spectrum.

We can now recognize the negative peaks of the second derivative spectrum of the protein lysozyme (Figure 7.3) as characteristic of each underlying Gaussian peak in the original spectrum. In our analysis of FT-IR spectra, the second derivative spectra are used to obtain the approximate number of component peaks. Initial guesses of the peak positions on the frequency axis are based on the positions of the negative second derivative peaks.

The second auxiliary technique used is Fourier deconvolution [2]. This partial resolution method was discussed in Section 4.A.3. The procedure is different from the Fourier transform used to convert the interferogram into a frequency-based specturm. Conceptually, Fourier deconvolution enhances the resolution of the spectrum by transforming the component Gaussian peaks into peaks with larger heights and smaller widths, while maintaining the same area and frequency as the original peaks.

A spectrum that has been partially resolved by Fourier deconvolution can be fit with the same model (Table 7.1) as used for the original spectrum. The requirement that both sets of data give identical parameters and peak areas can be used as a convergence criteria.

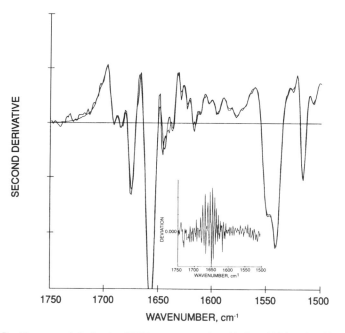

Figure 7.3 The second derivative FTIR spectrum of amide I and II bands of lysozyme in aqueous solution. The smooth line is experimental data. The jagged line on outer envelope is computed from the final regression analysis corresponding to the model and parameters found for the original spectrum, as described in the text. The inset shows the plot of connected residuals between calculated and experimental second derivative. (Reprinted with permission from [10], copyright by the American Chemical Society.)

Table 7.1 Model for the Analysis of Amide I and Amide II FT-IR Bands of Proteins

Assumptions:	Solvent background has been subtracted, water vapor and carbon dioxide removed

Regression equation:

$$A = A_o + \sum_{i=1}^{n} h_i \exp\{-(x - x_{0,i})^2 / 2W_i^2\}$$

Regression parameters:	*Data:*
$A_o\ h_i\ x_{0,i}\ W_i$ for $i = 1$ to n	A (absorbance) vs. x (frequency)
Auxiliary data sets:	Second derivative of spectrum, Fourier deconvoluted (FD) spectrum

Special instructions:

1. Use second derivative spectrum to identify number of peaks and their initial positions ($x_{0,i}$)
2. Enhance resolution of original spectrum using FD
3. Do initial regression analyses on the FD spectrum by first fixing the $x_{0,i}$. Take results of these initial fits and do regressions again to find all the parameters. Use these parameters to fit raw data spectrum
4. The major criteria for acceptance of the parameters are a final successful fit to the original spectrum using the fixed $x_{0,i}$ found from fitting FD spectrum, a random deviation plot, and SD $< e_y$. Confirm correct number of peaks (n) in model by using extra sum of squares F test (Section 3.C.1) for models with n, $n - 1$, and $n + 1$
5. True fractions of asparagine (ASN) and glutamine (GLN) should be subtracted from the fractional areas of bands at appropriate frequencies

B. Analysis of Spectra—Examples

B.1. Lysozyme

Determination of Component Bands   The protein spectrum is considered to be the sum of the individual absorption bands arising from specific structural components, such as α helices, β sheets, and turns. Fitting such a spectrum directly without prior knowledge of the unknown number of Gaussian bands would be a daunting task. To aid in this task, as mentioned previously, we first examine the second derivative of the spectrum (Figure 7.3). The negative peaks in the second derivative spectrum (cf. Figure 7.1) correspond to the component bands in the original spectrum. Inspection of the second derivative yields an estimate of the number of component bands and the approximate positions of these bands, which are used as initial guesses for the $x_{o,i}$.

The next step in the analysis is to enhance the resolution of the original spectrum by using the fourier deconvolution (FD) algorithm developed by Kauppinen *et al.* [2]. Care was taken to choose the correct values for the band width and resolution enhancement factor used in this algorithm, so that the FT-IR spectrum is not over- or underdeconvoluted. Underdeconvolution is recognized by the absence of a major band previously indicated by a negative peak in the second derivative spectrum. Overdeconvolution can be recognized by the appearance of large side lobes in the baseline region of the FD spectrum where no peaks should appear. This region should be flat, as in the original spectrum. As the deconvolution procedure progresses, analysis of the FD spectrum by nonlinear regression analysis can be used in an iterative fashion to help choose the FD parameters [9, 10].

The FD spectrum of lysozyme is shown in Figure 7.3. Nonlinear regression analysis is done on the FD spectrum using the model in Table 7.1, where A_o is a constant baseline offset. We first fix the set of $x_{0,i}$ at the initial values found from the second derivative spectrum. The regression analysis on the FD spectrum is now done to obtain estimates of component peak widths and heights. Starting with these values and the $x_{0,i}$ as initial parameter guesses, a second nonlinear regression is done on the FD spectrum in which all of the parameters are found. Finally, the parameters found in this second analysis are used in a nonlinear regression analysis on the original spectrum (raw data).

The FT-IR spectrum of lysozyme showing the amide I and amide II regions is shown in Figure 7.2. The outer envelope of the graph represents the experimental data. The underlying peaks represent the individual component bands making up the best fit to the model in Table 7.1, undertaken as discussed previously.

Quantitative criteria used to ensure a correct fit to the model are as follows:

1. Correlation of all the components bands and their positions with the negative second derivative peaks;
2. Agreement of FD and experimental baselines;
3. A SD $< e_y$ (instrumental noise);
4. A successful fit to the original spectrum using fixed $x_{0,i}$ and the other parameters found from the best fit of the FD spectrum.

In practice, attainment of these criteria may require several cycles of FD and regression, until an optimal fit is achieved.

Criterion 4 involves using the results of the regression analysis of the FD spectrum (Figure 7.2) to provide the number of bands and their frequencies, which are then fixed in a model to perform a nonlinear regression analysis of the original spectrum with the band widths and heights as parameters.

The final fit to the lysozyme FT-IR spectrum is illustrated by Figure 7.4 with its 29 component peaks. The inset shows that the residuals of the regression are reasonably random, indicating the model is a reliable fit to

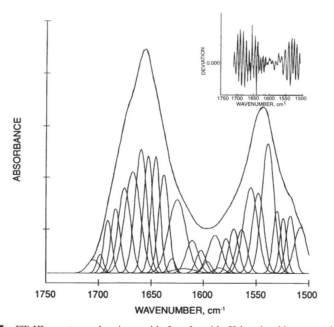

Figure 7.4 FT-IR spectrum showing amide I and amide II bands of lysozyme in aqueous solution. One outer envelope double line is the original spectrum. The second line on outer envelope is the computed best fit from nonlinear regression according to Table 7.1. The individual component bands underneath were constructed from the results of the nonlinear regression analysis. The inset shows plot of residuals (connected by a line) of the differences between the calculated and experimental absorbances vs. frequency. (Reprinted with permission from [10], copyright by the American Chemical Society.)

the data. Relative areas under the component bands of the original spectrum are in good agreement with those calculated from results of the regression analysis of the FD spectrum.

Further validation of the calculated components of the amide I and II bands can be obtained by comparing the second derivative FT-IR spectrum with the second derivative calculated from the model using the parameters found in the nonlinear regression. The results of such a comparison for lysozyme are shown in Figure 7.3. The inset of this figure shows a reasonably random residual plot, which further establishes the reliability of this methodology for quantitatively resolving FT-IR spectra of proteins into their individual component bands.

Component Band Assignments Now that the individual component bands of the lysozyme spectrum have been identified, their frequencies need to be assigned to the specific structural units that gave rise to them. The sum of the normalized areas under the peaks corresponding to a given structural feature can then be used to indicate the fraction of that particular feature in the protein [3, 4, 9, 10].

The band assignments given in Table 7.2 are based on theoretical considerations [5] as well as experimental correlations with X-ray crystal structures, as we shall demonstrate. The number of components bands found for a wide variety of proteins using the protocols outline in Table 7.1 are consistent with the numbers of bands predicted by theory [9, 10]. The assignments in Table 7.2 can be used for structural analysis based on the FT-IR of any protein in water.

Note that the assignments include bands for the side chains of asparagine (ASN) and glutamine (GLN) amino acid residues that occur within the amide I envelope [9, 10]. The true fractions of GLN and ASN obtained from the protein's amino acid content must be subtracted from the fractional areas of component bands at the appropriate frequencies. Any excess area may be assigned to a relevant secondary structural feature of the polypeptide backbone. If the fraction of area is less than the fraction of GLN and ASN present in the protein, then the experimental areas should be

Table 7.2 Secondary Structure Assignments of Amide I FT-IR Bands

Frequency, cm^{-1}	Assignment to structural unit
1681–1695	Turn
1673–1679	Turn, twisted sheet
1667–1669	GLN ($C = O$) and ASN ($C = O$) side chain, 3/10 helix, bent strand
1657–1661	α helix (A band)
1651–1653	GLN ($C - N$) and ASN ($C - N$) side chain, α helix (E band)
1643–1648	Disordered, irregular, gentle loop
1622–1638	Extended strand, rippled and pleated sheets

subtracted from the amide I envelope and all the remaining bands should be normalized.

Assignments of component bands in the amide I and amide II envelopes of lysozyme obtained from the nonlinear regression analysis are listed in Table 7.3. If this fractional area is less than the fraction of GLN and ASN residues present in the protein, then the experimental areas should be subtracted from the amide I envelope and all the remaining bands should be normalized so that their sum equals unity.

The band assignments in Table 7.3 can now be used to obtain the fractions of the different structural features. The area of each band is expressed as a fraction of the total band area of the amide I or amide II envelopes. From the relative fractional areas of these bands, we obtain estimates of the fractional amounts of the different structural features in the polypeptide chain. Replicate values of these fractions can be obtained from analysis of the amide I and amide II regions from the original and Fourier deconvoluted spectra, as shown in Table 7.4. Standard deviations suggest that the fractions for each feature are reproducible to $\pm 10\%$.

Lysozyme has a known crystal structure, and results from the FT-IR analysis can be compared to the X-ray crystal structure. The average FT-IR fractions agree reasonably well with the fractions from the crystallographic analysis. The agreement becomes a bit better when we realize that the X-ray data consider only one type of extended feature, but the FT-IR counts all such features. Also, the fraction of disordered regions is not measured directly by X-ray analysis but only as the difference between the sum of all other fractions and 1. Thus, the fraction of disordered regions will be an overestimate from the crystal structure, because the fraction of extended regions has been underestimated. Taking these discrepancies into account, we see that there is good agreement between the X-ray and FT-IR analyses (Table 7.4).

Table 7.3 Specific Secondary Structural Assignments in Analysis of Component FT-IR Bands of Lysozyme

Assignments	Helix	Extended (cm^{-1})	Disordered (loops)	Turns
Amide I	1660	1637	1646	1691
	1655	1629		1683
		1623		1675
				1668
Amide II	1540	1532	1548	1578
		1525		1571
				1564
				1556

Number of Component Peaks   Because of the large number of component peaks contributing to the FT-IR spectra of proteins, care must be taken that the correct number is used in the model. Careful attention must be paid to goodness of fit criteria (Section 3.C.1) as specifically outlined for the FT-IR analysis in Table 7.1. In addition, the *extra sum of squares F test* (Section 3.C.1) can be used as a statistical indicator to find the correct number of component peaks in the model.

An detailed study of the number of peaks needed to fit FT-IR spectra of proteins has been reported [9]. The number found in an analysis following the protocols in Table 7.1 is consistent with the number of peaks predicted by theory [5–8]. In general, it is found that inclusion of too many peaks in the model causes the insignificant component peaks to attain near zero or negative areas in the regression analysis. Inclusion of too few peaks causes the sum of squares of the deviations and the standard deviation of the regression to become larger. This latter situation is best evaluated with the extra sum of squares F test, since models with different numbers of peaks have different degrees of freedom.

B.2. Implications from Analysis of a Theoretical Spectrum of Lysozyme

Torii and Tasumi [6–8] calculated the theoretical FT-IR amide I spectrum from the three-dimensional X-ray crystal structure of lysozyme, using a Gaussian envelope of each peptide bond oscillator with a half-width at half-height of 3.0 cm^{-1}. This analysis predicted 16 peaks in the amide I envelope, although 2 of these peaks had very small fractional areas. The force constants used were optimized for D_2O and not H_2O. Therefore, we

Table 7.4   Fractions of Different Structural Features of Lysozyme from the FT-IR Regression Analysis Compared to Results from X-Ray Diffraction

Fractions:	Helix	Extended	Disordered	Turns
Source from Fourier deconvoluted spectrum				
Amide I	0.320	0.204	0.131	0.311
Amide II	0.278	0.175	0.183	0.364
Source from original spectrum				
Amide I	0.266	0.176	0.174	0.384
Amide II	0.323	0.176	0.090	0.411
Average ± SD	0.297 ± 0.029	0.183 ± 0.014	0.145 ± 0.043	0.365 ± 0.042
X-ray	0.310	0.155[a]	0.225[b]	0.310

[a] As fraction of β turns only, does not include all the extended features of the protein.
[b] Difference between 1 and the sum of all other reported fractions.

cannot directly compare the preceding analysis of experimental results for lysozyme with this theoretical spectrum, but we can analyze the published theoretical spectrum [6–8] into its component Gaussian peaks by the procedure in Table 7.1.

Figure 7.5 shows the Fourier deconvoluted theoretical FT-IR spectrum [6] with the best fit of a model consisting of the sum of 14 Gaussian bands. Attempts to use fewer than 14 bands resulted in poorer fits, as shown by the extra sum of squares F test. Addition of more peaks to the regression model caused the areas of the extra bands to approach a zero or negative values. Thus, the analysis by the protocols outlined in Table 7.1 identified all of the major component peaks in this theoretical spectrum.

We can compare these results with an analysis of the experimental spectrum of lysozyme in the amide I region only. The experimental Fourier deconvoluted FT-IR spectrum of lysozyme using an resolution enhancement factor (REF) of 3.8 is shown in Figure 7.6, with the graphical results of a regression analysis. Here, the amide I region is fit to the sum of 14 Gaussian peaks. A small standard deviation of the regression (within 0.1% of maximum A) and a random deviation plot were obtained. No additional bands could be successfully added to the 14 band model. Apparently, the two additional bands are too small to be detected. Good agreement of calculated and experimental data can be seen graphically. None of the

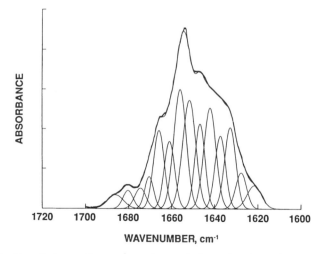

Figure 7.5 Best fit by nonlinear regression analysis to the theoretical amide I band of lysozyme. One outer envelope line is the theoretical spectrum; the second is the best fit to the model in Table 7.1. (Reprinted with permission from [9], copyright by the American Chemical Society.)

component bands were unacceptably broad, and the fit was similar to that when both amide I and II bands were analyzed (cf. Figures 7.2–7.4). Thus, the 14-peak model gave a successful best fit for both the theoretical and experimental amide I data for lysozyme.

Although the theoretical spectrum contained only an amide I band, the amide II band also appears in spectra obtained in water. We wish to point out a consequence of fitting only the amide I band in the spectrum. Attempts to fit the 14 bands to the experimental amide I envelope of the FD spectrum of lysozyme only, using a resolution enhancement factor of 2.5 ended with bands at the highest and lowest frequencies becoming unacceptably broad (Figure 7.7). This result indicates a problem, because the component bands arise from essentially the same type of vibrational mode in different environments and all band widths should be similar. The calculated and experimental outer envelopes of the amide I band show poorer agreement than in Figure 7.6. However, if the amide II and amide I bands are fit simultaneously (using REF of 2.5 and a model with 29 Gaussian components), inordinately large value bandwidths are not observed (see Figure 7.2) and the fit becomes acceptable. Therefore, it is important to use the amide II band with the amide I when analyzing the FD spectra generated using low REF values of 2–3. Calculations excluding the amide II envelope become very sensitive to the value of REF used, and REF values that are too small may lead to large errors in the estimated secondary structures.

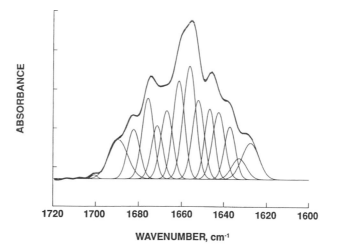

Figure 7.6 Best fit by nonlinear regression analysis of the Fourier deconvoluted lysozyme spectrum using a resolution enhancement factor (REF) of 3.8. Outer envelope lines as in Fig. 7.5. (Reprinted with permission from [9], copyright by the American Chemical Society.)

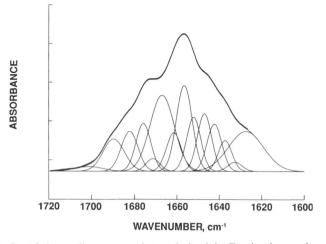

Figure 7.7 Best fit by nonlinear regression analysis of the Fourier deconvoluted lysozyme spectra using a resolution enhancement factor of 2.5. Outer envelope lines as in Fig. 7.5. (Reprinted with permission from [9], copyright by the American Chemical Society.)

B.3. Structural Comparisons with Crystal Structures of Proteins

A major advantage of FT-IR analyses for assessments of the secondary structure of proteins is that they can be done using only small amounts of protein in solutions [9, 10], in films [12], and in membranelike environments [12]. The classical method of determining structures of proteins is X-ray crystallography, which obviously requires a good protein crystal. In Table 7.4, we saw that the FT-IR analysis of the secondary structure of lysozyme gives a good correspondence with the X-ray structure. Further validation of the FT-IR method has been achieved by comparison of the FT-IR analyses of 14 proteins with their known crystal structures [10].

Computing fractions of structural features from X-ray crystal structures is somewhat imprecise [10]. Not only are fractions of α helix, turn, and extended conformations important, but the lengths of the helical and extended features as well as the presence of internal backbone hydrogen bonding may also contribute to the component bands of the amide I region. Thus, exact comparisons between FT-IR and X-ray crystallographic results will be subject to error. Also, the result may differ because of real differences between crystal structures and structures of the same protein in aqueous solutions. In other words, the X-ray results do not constitute flawless primary standards of comparison for the FT-IR results, but they are the best we have at present.

In the discussion that follows, the traditional Ramachandran plot calcu-

lated from the X-ray crystal structure was used to estimate the fractions
of structural features from data in the Brookhaven Protein Data Bank.
Details can be found in [9] and [10].

FT-IR and X-ray crystallographic comparisons for the 14 proteins and
polypeptides are summarized in Figures 7.8 to 7.11. Slopes of 1.0 and
correlation coefficients (r) of 1.0 in these plots would indicate perfect correlations. The FT-IR and X-ray results for the percent of helix (Figure 7.8)
are well correlated, with a slope of 0.91 and r close to 1. The scatter in this
plot increases below 20% helix. The correlations for the percent of turns
(Figure 7.9) and percent of extended conformations (Figure 7.10) are even
better, with r values of 0.99 and slopes close to 1. The correlation for the
percent disordered (Figure 7.11), is not as good as for the other structural

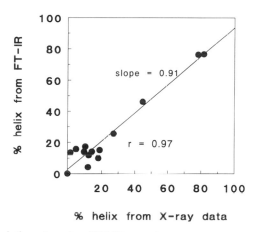

Figure 7.8 Correlation of results of FT-IR secondary structural analysis with X-ray crystal
structure for the percent of helical conformation. Proteins and % helix from FT-IR are:

Protein	% helix
Hemoglobin	76.7
Myoglobin	76.4
Cytochrome C	46.2
Lysozyme	25.6
Ribonuclease	10.0
Papain	15.2
P-tripsin inhibitor	4.2
α-Chymotrypsin	11.9
Trypsin	17.4
Elastase	14.1
Carbonic Anhydrase	14.2
β-Lactoglobulin	15.8
Conconavalin A	13.6
Oxytocin	0

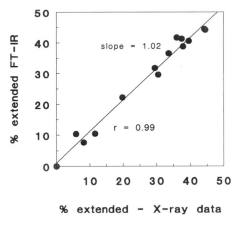

Figure 7.9 Correlation of results of FT-IR secondary structural analysis with X-ray crystal structure for the percent of extended conformation. Proteins are the same as in Figure 7.8.

features, with a smaller slope and *r* value. However, recall that the disordered features are not directly estimated by the X-ray analysis, so that error in both the X-ray and FT-IR data contribute to the scatter.

These results show overall good agreement of the FT-IR analysis with the X-ray crystal structures. Analyses of 14 proteins whose crystal structures are known showed agreement of secondary structure to within about 5% between the experimental FTIR and X-ray estimations. These correlations provide confidence in the reliability of the FT-IR secondary structural

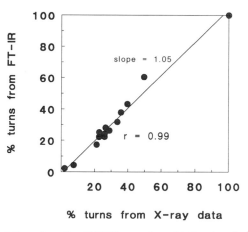

Figure 7.10 Correlation of results of FT-IR secondary structural analysis with X-ray crystal structure for the percent of turns and twists. Proteins are the same as in Figure 7.8.

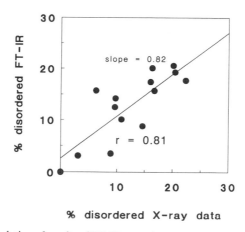

Figure 7.11 Correlation of results of FT-IR secondary structural analysis with X-ray crystal structure for the percent of disordered conformation. Proteins are the same as in Figure 7.8.

analysis. They suggest that such an analysis can provide an approximate picture of protein conformation in solutions. Large conformational changes, e.g., partial denaturation, can be readily identified by this type of analysis.

References

1. A. L. Lehninger, *Biochemistry*, 2nd ed. New York: Worth, 1976.
2. J. K. Kauppinen, D. J. Moffatt, H. H. Mantsch, and D. G. Cameron, "Fourier Self-deconvolution: A Method for Resolving Intrinsically Overlapped Bands," *Applied Spectroscopy* **32** (1981), p. 271.
3. H. Susi and D. M. Byler, "Resolution-Enhanced FT-IR Spectroscopy of Enzymes," *Methods. Enzymol.* **130** (1986), 290.
4. J. F. Rusling and T. F. Kumosinski, "New Advances in Computer Modeling of Chemical and Biochemical Data," *Intell. Instruments and Computers* (1992), pp. 139–145.
5. S. Krimm and J. Bandekar, "Vibrational Spectroscopy of Peptides and Proteins," *Adv. Protein Chem.* **38** (1986), pp. 183–363.
6. H. Torii and M. Tasumi, "Model Calculations on the Amide-I Infrared Bands of Globular Proteins," *J. Chem. Phys.* **96** (1992), pp. 3379–3387.
7. H. Torii and M. Tasumi, "Three-Dimensional Dorway–State Theory for Analyses of Absorption Bands of Many-Oscillator Systems," *J. Chem. Phys.* **97** (1992), pp. 86–91.
8. H. Torii and M. Tasumi, "Applications of Three-Dimensional Dorway–State Theory to Analyses of the Amide-I Infrared Bands of Globular Proteins," *J. Chem. Phys.* **97** (1992), pp. 92–98.
9. T. F. Kumosinski and J. J. Unruh, "FT-IR Spectroscopy of Globular Proteins in Water: Comparison with X-Ray Crystallographic Global Secondary Structure," *Talanta.*, in press.
10. T. F. Kumosinski and J. J. Unruh, "Global Secondary Structure Analysis of Proteins in Solutions: Resolution-Enhanced Deconvolution FT-IR Spectroscopy in Water," in T. F. Kumosinski and M. N. Leibman (Eds.), *Molecular Modeling*, ACS Symposium Series **576** (1994), pp. 71–99.

11. D. A. Skoog and J. J. Leary, *Principle of Instrumental Analysis,* 4th ed. New York: Saunders, 1992.

12. J. F. Rusling, A.-E. F. Nassar, and T. F. Kumosinski, "Spectroscopy and Molecular Modeling of Electrochemically Active Films of Myoglobin and Didodecyldimethyammonium Bromide," in T. F. Kumosinski and M. N. Leibman (Eds.), *Molecular Modeling,* ACS Symposium Series **576** (1994), pp. 250–269.

Chapter 8

Nuclear Magnetic Resonance Relaxation

A. Fundamentals of NMR Relaxation

Relaxation is a general term used to describe the movement of a system toward equilibrium after an injection of energy. When atomic nuclei in a molecule are excited in a pulsed nuclear magnetic resonance (NMR) experiment, the excited state may exist for a rather long time, often seconds to minutes [1]. Characteristic relaxation times estimated in an NMR experiment can be used to provide information about molecular motion. In the following paragraphs, we provide a simplified summary of pulsed NMR relaxation. The reader may consult texts on this subject for further details [1, 2].

At a magnetic field strength of zero, a diamagnetic sample is not magnetized. Suppose we now apply a magnetic field to this sample. A magnetic field will be induced in the sample, but this induced field will take some time to reach equilibrium. In the Bloch theory of NMR [1], the approach to equilibrium is assumed to be exponential. This assumption leads to the following expression for the time dependence of the magnetization M_z in the z direction, usually taken as the direction of the applied field:

$$M_z = M_0[1 + \exp(-t/T_1)] \tag{8.1}$$

where M_0 is the magnetization at thermal equilibrium, and T_1 is a characteristic relaxation time called the *longitudinal relaxation time*. *Longitudinal* denotes that the z component of the magnetization is of importance here. When the magnetization is moved away from the z direction, it will return to the z axis with a time constant T_1.

Different T_1 values may apply for the various atoms and nuclear environments in a molecule. In the limit of rapid isotropic motion in solution, the widths of the individual resonance peaks are inversely related to T_1 values.

We shall see that analysis of the NMR peak shape is one way to estimate T_1. Another way is by the so called inversion–recovery pulse sequence [1]. In this multiple-pulse NMR experiment, the z magnetization is inverted with a π pulse, so-named because it is π radians removed from the $+z$ axis. Thus, a π pulse is in the $-z$ direction. A variable time period is allowed for the magnetization to return partly to the z direction, and then a $\pi/2$ pulse is used for measurement [1].

Results from an inversion–recovery pulse sequence (Figure 8.1) show that the peaks at short measurement times are inverted, and at longer times they eventually return to the usual positive peaks. Nonlinear regression analysis of each peak height vs. time profile using the relevant model for peak height $A(t)$ at delay time t, provides the set of T_1 values characteristic of the different nuclear environments in the sample. The model is given as

$$A(t) = A_\infty[1 - 2\exp(-R_1 t)] \tag{8.2}$$

where A_∞ is the limiting peak height at long times and $R_1 = 1/T_1$ is the longitudinal relaxation rate.

As the z magnetization reappears with time constant T_1, the magnetization that has been induced in the x–y plane disappears. This process constitutes another relaxation. Called the *transverse relaxation,* it is assumed to be exponential and characterized by decay time T_2. If the only mechanism for transverse relaxation is the return of magnetization to the z direction, then $T_1 = T_2$. However, a number of factors can cause T_2 to be smaller

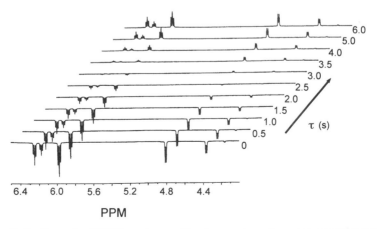

Figure 8.1 Example of NMR spectra resulting from an inversion–recovery pulse sequence. (Reprinted with permission from [1], copyright by Pergammon Press.)

than T_1. One is the inhomogeneity of the magnetic field, which depends on instrumental conditions and can be separated from other effects on T_2. Therefore, the symbol T_2 usually refers to longitudinal relaxation in the absence of inhomogeneous field effects.

Local magnetic fields characteristic of the sample may also cause T_2 to be smaller than T_1. Values of T_2 may provide information about molecules in the sample. In solids, for example, varying environments in the sample make T_2 very small, although T_1 may be very large bacause of the lack of motion [1]. We shall discuss examples of solid state NMR relaxation in Section C of this chapter.

The overall transverse relaxation time includes both T_2 and the influence of inhomogeneity of the magnetic field. This combined-effect relaxation is given the symbol T_2^*. It can be obtained from the free induction decay, which decreases exponentially with time constant T_2^*. However, the free induction decay is often more characteristic of the solvent [1]. Therefore, it is more reliable to obtain T_2^* from the width (W) of the NMR lines at half-height, using the relation

$$W = 1/\pi T_2^*. \tag{8.3}$$

Nonlinear regression onto the appropriate Gaussian, Lorentzian, or mixed Gaussian–Lorentzian model (see Section 3.B.2) can be used to obtain accurate estimates of the line width W and thereby T_2^*. According to theoretical predictions, the limiting line shape is Lorentzian when $T_1 = T_2$, and Gaussian when $T_1 \gg T_2$.

Spin-echo pulse NMR or spin-locking experiments [2] can also be used to obtain T_2. In the latter case, the model is a simple exponential decay in peak amplitude $A(t)$ at delay time t:

$$A(t) = A_0 \exp(-R_2 t) \tag{8.4}$$

where A_0 is the peak intensity at $t = 0$, and $R_2 = 1/T_2$ is the transverse relaxation rate.

B. Applications from NMR in Solution

B.1. NMR Relaxation of ^{2}D and ^{1}H in Aqueous Solutions of β-Lactoglobulin

Proton, deuterium, and oxygen-17 NMR relaxation can be used to investigate the amount and mobility of water bound to proteins. In this example, we illustrate the analysis of pulsed NMR relaxation data for ^{2}D and ^{1}H nuclei of water in a protein solution for such a system.

Figure 8.2 shows typical proton relaxation results for an aqueous solution of the protein β-lactoglobulin A. The transverse relaxation data gave a

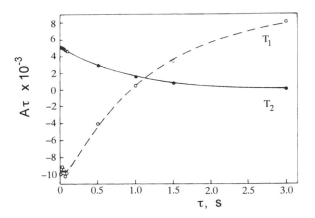

Figure 8.2 Proton NMR peak intensities vs. delay time τ for 0.061 g β-lactoglobulin A/g H_2O at pH 6.2, 30°C. Longitudinal relaxation data (T_1, o) from inversion recovery experiment with line representing best fit to single exponential model in eq. (8.2) and transverse relaxation data (T_2, ●) from spin-locking measurements with line representing best fit to sum of two exponentials in eq. (8.4). (Reprinted with permission from [3], copyright by Academic Press.)

good fit to a single exponential (eq. (8.4)), indicating a single relaxation rate with decay time T_2. However, the longitudinal relaxation data gave a best fit to the sum of two exponential terms ($n = 2$, Figure 8.2). This behavior is consistent with the finding that cross-relaxation between protons of water bound to the protein and protons on the protein itself contributes [4, 5] to the longitudinal relaxation rate.

Cross-relaxation is not significant for the *deuterium* NMR relaxation in solutions of proteins in D_2O. Models with a single time constant gave excellent fits for both longitudinal (Table 8.1, $n = 1$) and transverse relaxation (eq. (8.4)) of ^2D in solutions of β-lactoglobulin A in D_2O (Figure 8.3) [3].

Plots of relaxation rates in the D_2O solutions as $R_1 = 1/T_1$ and $R_2 = 1/T_2$ vs. concentration were linear up to 0.04 g β-lactoglobulin A per g D_2O.

Table 8.1 Model for the NMR Longitudinal Relaxation Rate

Assumptions: Inversion–recovery pulse sequence [1]
Regression equation:

$$A(t) = \sum_{i=1}^{n} A_{\infty,i}[1 - 2\exp(-R_{1,i}t)]$$

Regression parameters: *Data:*
  $A_{\infty,i}$ $R_{1,i}$ for $i = 1$ to n $A(t)$ vs. t
Special instructions: Begin testing models with $n = 1$. Test for best fit by comparing residual plots and using the extra sum of squares F test (Section 3.C.1)

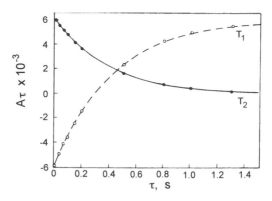

Figure 8.3 Deuterium NMR peak intensities vs. delay time τ for 0.029 g β-lactoglobulin A/g D_2O at pH 6.2, 30°C. Longitudinal relaxation data (T_1, o) from inversion recovery experiment with line representing best fit to single exponential model in eq. (8.2) and transverse relaxation data (T_2, ●) from spin-locking measurements with line representing best fit to single exponential in eq. (8.4). Results indicate no cross-relation with protein (Reprinted with permission from [3], copyright by Academic Press.)

These results were used to obtain the correlation time of D_2O bound to the protein [3]. At 30°C and pH 6.2, this correlation time was 10 ns, which was used to obtain a hydrodynamic radius of the protein of 23 Å at this pH. At pH 4.65, a correlation time of 22.5 ns translates into a radius of 30 Å indicative of association of the protein. Thermodynamic parameters were also obtained by methods outlined in the original paper [3].

B.2. Line Shape Analysis—β-Lactoglobulin with Deuterated Lysines

As mentioned in Section A, analysis of NMR line shapes and estimation of peak width W can provide values of T_2 via eq. (8.3). In this section we shall see an example of the estimation of correlation times of a deuterium-labeled protein from the peak width.

If observed line shapes for the ^2D NMR spectra are Lorentzian, the effective correlation time ($\tau_{2,\text{eff}}$) for [$^2D_6$]isopropyl groups incorporated into the protein can be calculated directly. Under such conditions, effective correlation times are calculated from the relationship [6]

$$\tau_{c,\text{eff}} = \frac{(8/3)\pi W}{(e^2qQ/h)^2} \qquad (8.5)$$

where (e^2qQ/h) = quadrupole coupling constant and W is the line width at half-height in Hz. The quadrupole coupling constant in our example is 170 kHz.

Deuterium labeling was achieved by covalently binding [$^2D_6$]isopropyl groups to 80% of the lysine residues of β-lactoglobulin [6]. Figure 8.4A shows the 61.25 MHz spectrum of the deuterium-labeled β-lactoglobulin in an aqueous buffer. Underlying the raw data envelope are the two Lorentzian peaks (see Section 3.B.2 for models) found to give the best fit by a program employing a Gauss–Newton algorithm. The resulting residual plot is shown at the bottom. The corresponding spectrum of deuterium-labeled β-lactoglobulin in 6 M guanidine hydrochloride is shown in Figure 8.4B. Models with one or two Gaussian peaks or with a single Lorentzian peak resulted in poor fits with clearly nonrandom residual plots.

Thus, a model composed of the sum of two Lorentzian peaks with nearly identical positions but different linewidths produced the best fit for spectra of the intact protein at 61.25 MHz in either solution. Using the results of nonlinear regression analyses, residual plots, and F tests confirmed that the two Lorentzian model gave the best fit.

Two populations of lysine residues were apparent in spectra obtained in either aqueous buffer or 6 M guanidine hydrochloride. Slow and fast correlation times were determined from the individual linewidths obtained from the nonlinear regression analysis of the spectra. The numbers of residues corresponding to each $\tau_{c,\mathrm{eff}}$ (eq. (8.5)) were obtained from the ratio of the areas of the two Lorentzian curves. In an aqueous buffer, the equivalent of 9.7 modified lysine residues were observed. Here, 6.5 residues

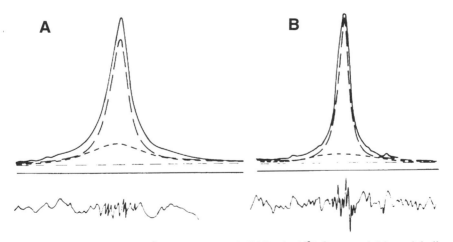

Figure 8.4 (A) 61.25 MHz ^2D NMR spectra (solid lines) of [$^2D_6$]isopropyl-β-lactoglobulin in tris buffer pH 7.5 and (B) in 6 M guanidine hydrochloride, pH 7.5. Each spectrum gave the best fit by nonlinear regression to a model composed of the sum of two Lorentzian peaks (see Section 3.C.2), shown underlying the spectra. Traces representing residual plots are shown at the bottom, below each spectrum. (Reprinted with permission from [6], copyright by the American Chemical Society.)

had a fast average correlation time ($\tau_{c,\text{eff}}$ = 70 ps), 3.2 residues had a slow, but observable average correlation time ($\tau_{c,\text{eff}}$ = 320 ps), and the remaining 2.3 residues had a $\tau_{c,\text{eff}}$ apparently so long that they could not be resolved from the baseline at this field strength. In 6 M guanidine hydrochloride, where all of the modified lysines were observed, the equivalent of 8.7 residues had a fast average correlation time ($\tau_{c,\text{eff}}$ = 20 ps) and 3.3 residues a slow average correlation time ($\tau_{c,\text{eff}}$ = 320 ps).

The best fit for the spectrum of deuterium-labeled β-lactoglobulin after hydrolysis to individual amino acids (not shown) was a single Lorentzian. This result showed that in this hydrolyzed sample, all [$^2D_6$]isopropyl groups were equally mobile.

B.3. Influence of Frequency on Deuterium Relaxation in D$_2$O Solutions of Proteins

B.3.a. Field Dispersion in Protein Solutions

NMR relaxation provides additional molecular information if it can be done at a series of frequencies. The general observation in NMR frequency dependence experiments in solutions of proteins is that the longitudinal NMR relaxation rates ($R_1 = 1/T_1$) of ^2D nuclei are independent of the field at very low NMR frequencies. As NMR frequency (and magnetic field strength) increases, R_1 gradually decreases in a sigmoid-shaped curve until finally it becomes independent of the frequency and field strength [7]. An example, shown in Figure 8.5 illustrates the *field dispersion* of the NMR relaxation.

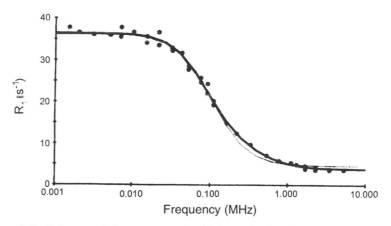

Figure 8.5 Influence of frequency on the ^2D-NMR longitudinal relaxation rate for a 30.6 mg mL^{-1} solution of α-hemocyanin in 80% D$_2$O/20% H$_2$O. The solid curve represents the best fit of the model in Table 8.3 with two correlation times (i.e., k = 2), with $t_{c,1}$ = 180 ns and $t_{c,2}$ = 1070 ns. (Reprinted with permission from [7], copyright by Elsevier.)

The limiting high-frequency value for R_2 of water bound to proteins is larger than the frequency-independent relaxation rate of pure water. The inflection point of the sigmoid curve, that is, the midpoint between the high and low frequency plateaus, is related to the protein rotational correlation time [8–10]. Field dispersion data fully describe the spectral density function(s) that characterize the various dynamic processes in the system responsible for NMR relaxation.

In interpreting NMR relaxation data for protein solutions, it is instructive to have an idea of the time scale of the dynamic processes involving water. For a globular protein, water molecules are generally thought to exchange between a "free" (bulk water) state and a "bound" state with a approximate lifetime of 10^{-5}–10^{-6} [11]. Water molecules bound to the protein reorient with the protein, with correlation times on the order of 10^{-8}–10^{-9}s, and diffuse along the surface of the protein on a time scale of 10^{-7} s [11]. Other possible dynamic processes that involve "bound" water molecules include the motion of hydrated protein side chains (10^{-10}–10^{-11} s), rotation of water molecules around their bonding axis to the protein surface (10^{-11} s), and the *chemical* exchanges of hydrogen between water molecules and ionizable protein side chains as well as between "bound" and "free" water molecules (10^{-4} s) [7–11].

These time scales can be compared to the translation/rotation of "free" water (10^{-12} s) and proton exchange between 'free' water molecules with a lifetime on the order of 10^{-4} s [11]. The contribution of NMR relaxation due to motions faster than 10^{-10} s is frequency independent in the experimentally accessible frequency range. Motions involving water that are slower than this limit give rise to the frequency dependence of NMR relaxation in protein solutions.

B.3.b. Models for Deuterium Relaxation in Protein Solutions

The simplest model of protein hydration is a two-state one, where water (i.e., D_2O) exchanges chemically between a bound and a free state. Using the general approach to equilibrium problems described in Chapter 6 and making the reasonable assumption that T_1 for bound water is much less than that of free water, the observed longitudinal relaxation time is [7]

$$\frac{1}{T_{1,obs}} = \frac{f_1}{T_{1,1} + \tau_m} + \frac{f_0}{T_{1,0}} \tag{8.6}$$

where the symbols are defined as follows:

f_1 = fraction of bound water,
f_0 = fraction of free water,
$T_{1,1}$ = relaxation time of ^2D for bound water,
$T_{1,0}$ = relaxation time of ^2D for free water,
τ_m = lifetime of a water molecule in the bound state.

Because $T_{1,1} \gg \tau_m$, eq. (8.6) can be written as

$$R_{1,\text{obs}}(\omega) = f_1 R_{1,1}(\omega) + f_0 R_{1,0} \tag{8.7}$$

where the relaxation times have been replaced by the corresponding relaxation rates ($R_1 = 1/T_1$). The frequency (ω)-dependent relaxation rates $R_{1,\text{obs}}$ and $R_{1,1}$ are distinguished from the frequency independent $R_{1,0}$ ones for free water. In other words, the frequency dependent component of the measured deuterium relaxation rate is directly proportional to the longitudinal relaxation rate of the deuteriums in water bound to the protein.

Three independent dynamic processes with separate correlation times (τ) need to be considered for the "bound" water molecules. These are radial diffusion perpendicular to the protein surface involving "bound"–"free" chemical exchange (τ_{rad}), lateral diffusion parallel to the protein surface (τ_{lat}), and rotation of the protein itself (τ_{rot}) [7]. Then, the correlation time associated with the nanosecond time scale motion of the protein-bound water (τ_c) is given by

$$\frac{1}{\tau_c} = \frac{1}{\tau_{\text{rad}}} + \frac{1}{\tau_{\text{lat}}} + \frac{1}{\tau_{\text{rot}}}. \tag{8.8}$$

A consequence of eq. (8.8) is that the fastest of the dynamic processes dominates the observed field dispersion of the relaxation; that is, it governs $1/\tau_c$. For a moderately sized globular protein $\tau_{\text{rot}} \sim 10^{-8}$ s $< \tau_{\text{lat}} \sim 10^{-7}$ s $< \tau_{\text{rad}} \sim 10^{-6}$ s. Thus, τ_c is approximately the same as τ_{rot}. As expected from this relation, an increase of τ_c is observed with increasing protein size [7].

Models for the influence of frequency (ω) on the longitudinal relaxation rate were derived from the theory of quadrupolar relaxation [12]. The model is outlined in Table 8.2 for a system with a single correlation time. This model holds for a solution of monomeric protein in D_2O, characterized by a single correlation time τ_c.

Solutions of protein in equilibrium with their aggregates are characterized by more than one correlation time. Each distinct jth aggregate has its own correlation time $\tau_{c,j}$. The general model for the frequency dependence of R_1 that accounts for such aggregates is given in Table 8.3.

Table 8.2 Model for Frequency Dependence of the NMR Longitudinal Relaxation Rate

---

Assumptions: Single correlation time, τ_c
Regression equations:

$$R_1 = A + BF(\omega, \tau_c)$$

$$F(\omega,\tau_c) = \frac{0.2\tau_c}{1 + \omega^2\tau_c^2} + \frac{0.8\tau_c}{1 + 4\omega^2\tau_c^2}$$

Regression parameters: *Data:*
  A B τ_c R_1 vs. ω
Special instructions: Also test models with more than one τ_c (see Table 8.3)

Table 8.3   General Model for Frequency Dependence of NMR Longitudinal Relaxation
for Aggregating Protein Solutions

Assumptions: Multiple correlation time, $\tau_{c,j}$
Regression equations:

$$R_1 = A + \sum_{j=1}^{k} B_j F(\omega, \tau_{c,j})$$

$$F(\omega, \tau_{c,j}) = \frac{0.2\tau_{c,j}}{1 + \omega^2\tau_{c,j}^2} + \frac{0.8\tau_{c,j}}{1 + 4\omega^2\tau_{c,j}^2}$$

Regression parameters: *Data:*
  A B_j $\tau_{c,j}$ for $j = 1$ to k R_1 vs. ω
Special instructions: Test for goodness of fit by comparing residual plots and using the
  extra sum of squares F test (see Section 3.C.1)

B.3.c. Analysis of Field Dispersion Results for Various Proteins

The models in Tables 8.2 and 8.3 were applied to data on the frequency
dependence of ^{2}D-NMR for solutions of the proteins lysozyme, bovine
serum albumin (BSA), alkaline phosphatase, and hemocyanin. In all cases,
models with at least two independent correlation times gave the best fit to
the data. Some typical results are illustrated graphically in Figures 8.5 and
8.6. For globular proteins, the shortest correlation time (10^{-8} s) was identified with the rotational correlation time of the monomeric protein. A second
correlation time (10^{-7} s) was attributed to aggregates of the proteins.

Extra sum of squares F tests confirmed that more than a single correlation
time was required to fit the data with a 99% probability of best fit in all
cases. Lysozyme, phosphatase, 10% BSA, and hemocyanin data required
two correlation times (Table 8.3, $k = 2$), whereas three were required for
25% and 33% BSA ($k = 3$). The latter behavior is indicative of increased
aggregation as protein concentration is increased. The B_j parameters in the
model can be used to estimate the amount of hydration [7].

In the past, field dispersion data for protein solutions have been fitted
by the Cole–Cole equation [5]. However, this expression has little a priori
validity. The correlation time found from the Cole–Cole analysis are overes-
timates compared to those found by analysis using the models in Table
8.3 [7].

B.4. NMR Relaxation Studies of the Hydration of Soy Protein

B.4.a. ^{17}O NMR

NMR relaxation of ^{17}O is free from the complications of cross-relaxation
and chemical exchange of protons that influence relaxation of ^{1}H and ^{2}D.
Thus, ^{17}O relaxation provides a direct probe of the hydration of proteins

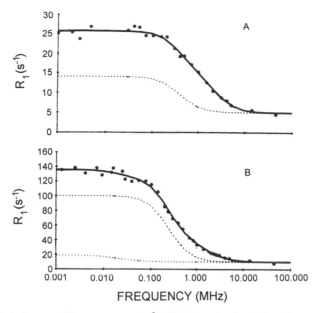

Figure 8.6 Influence of frequency on the ^{2}D-NMR longitudinal relaxation rate at 18°C for (A) 10% BSA solution in D_2O; solid line is best fit of the model with two correlation times $t_{c,1} = 40$ ns and $t_{c,2} = 228$ ns; (B) 33% BSA solution in D_2O; solid line is best fit of the model with three correlation times $t_{c,1} = 48$ ns, $t_{c,2} = 360$ ns, and $t_{c,3} = 4400$ ns. Dotted lines show major components of the relaxation. (Reprinted with permission from [7], copyright by Elsevier.)

[8,13]. The NMR relaxation rate of ^{17}O in water increases when the water is bound to macromolecules because of the asymmetric electrostatic interactions at the binding site and the longer effective correlation times of bound water. On the other hand, reorientation of water at the binding site and fast exchange between bound and free water tend to decrease the ^{17}O NMR relaxation rate.

Starting from the two state model, that is, bound and free water, in eq. (8.6), a linear model was derived for the dependence of the relaxation rates on protein concentration (C_p):

$$R_{m,\text{obs}} = n_H(R_{m,1} - R_{m,0})C_p + R_{m,0} \qquad (8.9)$$

where the m denotes either the longitudinal (R_1) or transverse relaxation rate (R_2), $R_{m,1}$ is the relaxation rate for bound water, $R_{m,0}$ is the relaxation rate for free water, C_p is the concentration of protein in g/g water, and n_H is the hydration number in g water bound/g protein. If the correlation time of bound water can be estimated or calculated [13], the $R_{m,1}$ value can be found. Then, the slope of a plot of $R_{m,\text{obs}}$ vs. C_p can be used to obtain n_H.

Inversion–recovery peak heights vs. time for soy protein gave a good fit to the model in Table 8.1 with $n = 1$. Typical results are shown in Figure

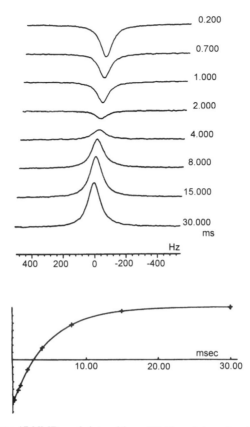

Figure 8.7 Oxygen-17 NMR peak intensities of D_2O vs. interpulse delay in an inversion recovery experiment to determined T_1: (Top) Spectra for a 2% (w/w) dispersion of soy protein in D_2O at 21°C and effective pH in D_2O (pD) = 7.4; (Bottom) Line is the best fit of the peak heights vs. time to model in eq 8.2, (+) are experimental data. (Reprinted with permission from [13], copyright by the American Chemical Society.)

8.7. The ^{17}O relaxation rates were linear with protein concentration (Figure 8.8), and the slopes of these lines were used to obtain an n_H of 0.33 g water/ g protein [13]. This value agrees with estimates from proton NMR at $-50°C$ and from differential scanning calorimetry.

B.4.b. ^1H NMR Relaxation of Soybean Protein

As mentioned in Section B.1, cross-relaxation between protons on water and the protein can cause problems in the analysis of ^1H NMR relaxation data. In the presence of cross-relaxation, the following equation can be used for the dependence of relaxation rates on protein concentration:

$$R_{m,\text{obs}} = [n_H(R_{m,1} - R_{m,0})C_p + R_x] + R_{m,0} \qquad (8.10)$$

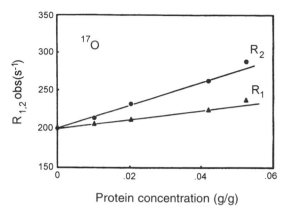

Figure 8.8 Influence of protein concentration the longitudinal (▲) and transverse (●) ^{17}O NMR relaxation rates for dispersions of soybean protein in D_2O at 21°C and neutral pH. (Reprinted with permission from [13], copyright by the American Chemical Society.)

where R_x is the cross-relaxation rate. The influence of cross relaxation can often be neglected for $R_{2,obs}$, but its dependence on protein concentration is nonlinear. By neglecting R_x and replacing C_p by the chemical activity of the protein as expressed by a virial expansion [13], the model for $R_{2,obs}$ becomes

$$R_{2,obs} = R_{m,0} + [n_H(R_{2,1} - R_{2,0})]$$
$$[C_p \exp(2B_0C_p + 2B_{0.5}C_p^{0.5} + 0.667B_{1.5}C_p^{1.5} + 1.5B_2C_p^2 + \ldots)]. \qquad (8.11)$$

The parameters B in eq. (8.11) are called *virial coefficients* and can be interpreted in terms of various molecular interactions [13].

The dependence of $R_{2,obs}$ on soy protein concentration gave good fits to eq. (8.11) (Figure 8.9) [13]. In the range of protein concentrations examined, all the data at pH $\leq$ 9.1 gave best fits to the model using only the B_0 virial coefficient. Addition of more terms to the model did not provide a statistically better fit, as was shown by using the extra sum of squares F test. Inclusion of the B_0 and the B_2 terms was required to give the best fit to the pH 11 data.

C. Applications from NMR in the Solid State

C.1. Introduction

Rapid random molecular motion in liquids usually averages chemical shift anisotropies and J couplings in NMR to discrete isotropic values. Dipolar and quadrupolar interactions are averaged nearly to zero, leaving finite linewidths in solutions that are caused mainly by the inhomogeneity

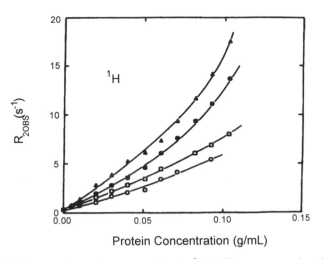

Figure 8.9   Influence of protein concentration the ^{1}H NMR transverse relaxation rates for dispersions of soybean protein in D_2O at 22°C and, in ascending order, pH 4.5, 7.0, 9.1, and 11.0. Solid lines are the best fits to eq. (8.11), after tests with different numbers of parameters. The best fits were obtained using parameters $R_{m,0}$, n_H, C_p, and B_0 for pH 4.5–9.1, and parameters $R_{m,0}$, n_H, C_p, B_2, and B_0 for pH 11. (Reprinted with permission from [13], copyright by the American Chemical Society.)

of the magnetic field. For samples in the solid state, all of these anisotropic interactions are manifested, to the detriment of the NMR signals.

With the development of cross polarization/magic angle spinning (CP/MAS) experiments and high-powered decoupling, solid state NMR has become an important technique for characterizing the bulk structures of solids [14, 15]. Magic angle spinning simulates motional averaging for the solid, and high-power decoupling minimizes dipolar interactions. Using cross-polarization, a gain in sensitivity for dilute nuclear spin systems such as ^{13}C is achieved via dipolar coupling to the abundant ^{1}H spin system.

C.2. Analysis of Line Shapes

Chemical shifts determined by CP/MAS experiments have been used to characterize the secondary structure of solid polypeptides [16]. In this section, we extend the NMR line shape analysis from Section B.2 to the structural investigation of solid proteins and polypeptides.

Solution NMR lines tend toward Lorentzian shapes in the limit where $T_1 = T_2$. Restricted molecular motion in the solid state gives rise to a statistical ensemble of Lorentzian lines for a given type of nucleus. The resulting broad line has Gaussian nature. Therefore, powdered samples characterized by little intermolecular motion should have Gaussian line shapes. Proteins and other polymers that might be expected to have some degree of molecular motion even in the solid will feature CP/MAS peaks with a fraction of Lorentzian character roughly proportional to the degree of motion.

Nonlinear regression analysis of CP/MAS data onto the appropriate Gaussian, Lorentzian, or mixed Gaussian–Lorentzian model (Section 3.B.2 and Table 3.8) yields the fraction of Lorentzian character of the peaks. An increased amount of Lorentzian character suggested increased local mobility of protein and polypeptide subunits.

Solid adamantane was investigated first as a standard. This small rigid molecule should have pure Gaussian shapes for both of its CP/MAS ^{13}C lines. Indeed the all-Gaussian model gave a better fit than the Gaussian–Lorentzian model in nonlinear regression analysis. Peak areas were within 1% of values computed from the number of carbons in the molecule.

Figure 8.10 shows the CP/MAS ^{13}C spectrum and nonlinear regression results for glutamic acid, and Figure 8.11 shows the results for polyglutamic

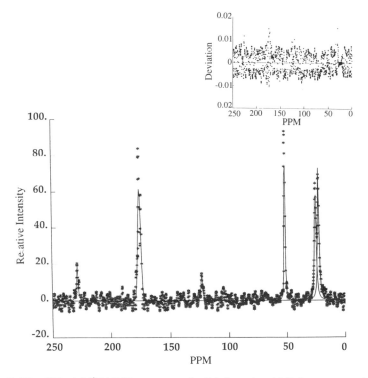

Figure 8.10 CP/MAS ^{13}C NMR spectrum of solid glutamic acid. Points are experimental, and outer envelope line represents the best fit by nonlinear regression onto a multiple Gaussian–Lorentzian peak model (Table 3.8). Underlying peaks are the components computed from results of the nonlinear regression analysis. (The authors thank Dr. M. Alaimo for the original data.)

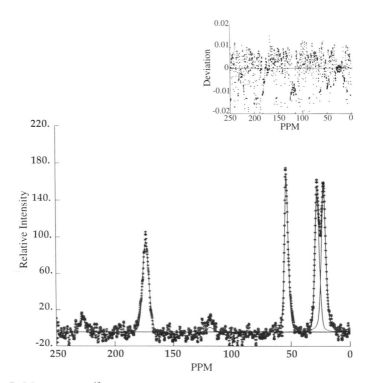

Figure 8.11 CP/MAS ^{13}C NMR spectrum of powdered polyglutamic acid. Points are experimental, and outer envelope line represents the best fit by nonlinear regression onto a multiple Gaussian–Lorentzian peak model (Table 3.8). Underlying peaks are the components computed from results of the nonlinear regression analysis. (The authors thank Dr. M. Alaimo for the original data.)

acid. The carbonyl resonances at about 170–175 ppm are approximately 80% Gaussian for both samples when using nonlinear regression onto the Gaussian–Lorentzian model (Table 3.8). One problem that can be seen in fitting the very narrow lines of glutamic acid is that a limited number of data points are available. This is probably responsible for the errors seen at the maxima of the peaks in the glutamic acid spectrum. Polyglutamic acid, with broader peaks, has more data points and shows better agreement between the data and the model at the peak..

The methyl and methylene peaks in the 5–70 ppm regions of the glutamic acid and polyglutamic acid spectra were predominantly Gaussian. Peaks for the polypeptide are significantly broader than for the monomeric glutamic acid. This broadening can be attributed to motion of the backbone secondary structure of polyglutamic acid.

Nonlinear regression analysis of the CP/MAS ^{13}C spectrum of the protein lysozyme (Figure 8.12) revealed different motion characteristics from poly-

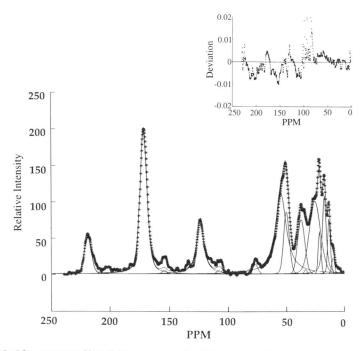

Figure 8.12 CP/MAS ^{13}C NMR spectrum of solid lysozyme. Points are experimental, and outer envelope line represents the best fit by nonlinear regression onto a multiple Gaussian–Lorentzian peak model (Table 3.8). Underlying peaks are the components computed from results of the nonlinear regression analysis. (The authors thank Dr. M, Alaimo for the original data.)

glutamic acid. Lysozyme has a carbonyl resonance with 50% Lorentzian character, indicating a rather high degree of motion. The methyl and methylene peaks were largely Gaussian.

References

1. A. Derome, *Modern NMR Techniques for Chemistry Research.* Elmsford, N.Y.: Pergammon Press, 1987.
2. E. D. Becker, *High Resolution NMR.* New York: Academic Press, 1980.
3. T. F. Kumosinski and H. Pessen, "A Deuteron and Proton NMR Relaxation Study of β-Lactoglobulin A Association," *Arch. Biochem. and Biophys.* **218** (1982), pp. 286–302.
4. H. T. Edzes and E. T. Samulski, "The Measurement of Cross-Relaxation Effects in the Proton NMR Spin-Lattice Relaxation of Water in Biological Systems: Hydrated Collagen and Muscle," *J. Magn. Reson.* **31** (1978), pp. 207–229.
5. S. H. Koenig, R. G. Byant, K. Hallenga, and G. S. Jacob, "Magnetic Cross-Relaxation among Protons in Protein Solutions," *Biochemistry* **17** (1978), pp. 4348–4358.

6. E. M. Brown, P. E. Pfeffer, T. F. Kumosinski, and R. Greenberg, "Accessibility and Mobility of Lysine Residues in β-Lactoglobulin," *Biochemistry* **27** (1988), pp. 5601–5610.

7. L. T. Kakalis and T. F. Kumosinski, "Dynamics of Water in Protein Solutions: The Field Dispersion of Deuterium NMR Longitudinal Relaxation," *Biophysical Chem.* **43** (1992), pp. 39–49.

8. B. Halle, T. Anderson, S. Forsen, and B. Lindman, "Protein Hydration from Water Oxygen-17 Magnetic Relaxation," *J. Am. Chem. Soc.* **103** (1981), pp. 500–508.

9. S. H. Koenig, K. Hallenga, and M. Shporer, "Protein-water interactions studied by Solvent ^{1}H, ^{2}H, and ^{17}O Magnetic Relaxation," *Proc. Natl. Acad. Sci. USA* **72** (1975), pp. 2667–2671.

10. K. Hallenga and S. H. Koenig, "Protein Rotational Relaxation as Studied by Solvent ^{1}H and ^{2}H Magnetic Relaxation," *Biochemistry* **15** (1976), pp. 4255–4264.

11. K. J. Packer, "The Dynamics of Water in Heterogeneous Systems," *Phil. Trans. R. Soc. London B* **278** (1977), pp. 59–87.

12. A. Abragam, *Principles of Nuclear Magnetism*, pp. 264–353. New York: Oxford University Press, 1961.

13. L. T. Kakalis, I. C. Baianu, and T. F. Kumosinski, "Oxygen-17 and Proton Magnetic Relaxation Measurements of Soy Protein Hydration and Protein–Protein Interactions in Solution," *J. Ag. Food Chem.* **38** (1990), pp. 639–647.

14. A. Pines, M. G. Gibby, and J. S. Waugh, "Proton-Enhanced Nuclear Induction Spectroscopy. A Method for High Resolution NMR of Dilute Spins in Solids," *J. Chem. Phys.* **56** (1972), pp. 1776–1777.

15. J. Shaefer, E. A. Stejskal, and R. Buchdahl, "Magic-Angle 13C NMR Analysis of Motion in Solid Glassy Polymers," *Macromolecules* **10** (1977), pp. 384–405.

16. H. R. Kricheldorf and D. Muller, "Secondary Structure of Peptides. 3. 13C NMR Cross Polarization/Magic Angle Spinning Spectroscopic Characterization of Solid Polypeptides," *Macromolecules* **16** (1983), pp. 615–623.

Chapter 9

Small-Angle X-Ray Scattering (SAXS) of Proteins

A. Theoretical Considerations

Scattering of X-rays at small angles can be used to obtain parameters such as the hydrated volume, the external surface area, the electron density, and the degree of hydration of proteins and other macromolecules in solution. The basic theory has been presented by Luzzati, Witz, and Nicolaieff [1] and Van Holde [2].

X-ray scattering measurements can be made at angles (θ) small enough to allow a good extrapolation to $\theta = 0$. The molecular weight of a solute can be estimated from this extrapolated scattering [2]. The intensity of X-ray scattering at a distance r from a single electron is

$$i_e(\theta) = \frac{I_0}{r^2}\left(\frac{e^2}{mc^2}\right)^2\left(\frac{1 + \cos^2\theta}{2}\right) \tag{9.1}$$

where e is the charge on an electron and m is its mass, I_0 is the X-ray source intensity, and c is the speed of light. For N molecules, each with n electrons, in a cubic centimeter of solution, at $\theta = 0$,

$$i(0) = \frac{I_0}{r^2}\left(\frac{e^2}{mc^2}\right)^2 n^2 N. \tag{9.2}$$

For experiments in solution, n is replaced by $(n - n_0)$, representing the excess electrons in a solute molecule over the volume of solvent it displaces. Recasting eq. (9.2) in terms of molecular weight (M), the weight concentration of solute (C) and Avogadro's number (N_A), we have [2]

$$i(0) = \frac{I_0}{r^2}\left(\frac{e^2}{mc^2}\right)^2 N_A \left(\frac{n - n_0}{M}\right)^2 MC. \tag{9.3}$$

The quantity $(n - n_0/M)$ is essentially a ratio of atomic number to atomic weight and depends on the nature of the macromolecule. Thus, the scattering intensity depends on of the types of molecules doing the scattering and may be analyzed to obtain a variety of useful molecular parameters.

To obtain the radius of gyration (R_G) of the macromolecule, the angular dependence of the X-ray scattering at small angles is used. The equations and the notation that follow [3] apply to globular particles and to so-called infinite slit collimation. These experimental conditions must be satisfied for the systems under examination.

The excess scattered intensity $j_n(s)$ is the scattered intensity of the sample normalized with respect to the intensity of the incident beam and corrected for the scattering of the blank. It is related to the scattering angle (θ) by [3]

$$j_n(s) = j_n(0) \exp[-(4/3)\pi^2 R_a^2 s^2] + \phi(s) \tag{9.4}$$

where the symbols are as follows:

$s = (2 \sin \theta)/\lambda$,
θ = scattering angle,
λ = wavelength of the source radiation (e.g., 1.542Å for the Cu-K$_\alpha$ doublet),
$j_n(0)$ = normalized intensity extrapolated to $\theta = 0$,

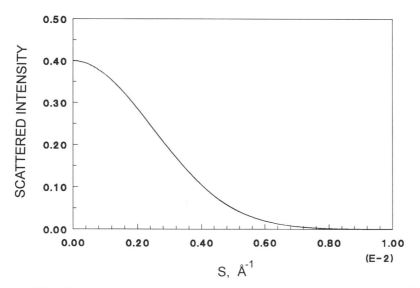

Figure 9.1 Theoretical scattering intensity curve according to eq. (9.4) for $(4/3)\pi^2 R_a^2 = 1/0.000012$, $j_n(0) = 0.4$, and $\phi(s) = 0$.

Table 9.1 General Model for Analysis of SAXS Data

Assumptions: Solvent or buffer background has been subtracted. This can be done by
fitting background curves to the model that follows with $p = 2$, then subtracting the
fitting equation from the raw data [4]
Regression equation:

$$j_n(s) = b_0 + \sum_{i=1}^{p} j_{n,i}(0) \exp\{-(4/3)^2 \pi^2 R_{a,i}^2 s^2\}$$

$$s = 2 \sin \theta / \lambda; \qquad b_0 = \text{baseline offset}$$

Regression parameters: *Data:*
b_0 and $j_{n,i}(0)$; $R_{a,i}$ for $i = 1$ to p $j_{n,i}(0)$ vs. s
Special instructions: Fit data to $p = 1$, $p = 2$, $p = 3$. . . . , models in succession.
Test for goodness of fit by comparing residual plots and by using the extra sum of squares
F test (see Section 2.C.1); use proper weighting, such as for a counting detector $w_j = 1/y_j$
in eq. (2.9) (cf. Section 2.B.5)

Table 9.2 Parameters Available from SAXS Data [1]

Symbol	Parameter	Equation
m_{app}	Apparent molecular mass, in e-/molecule, at concentration c_e	$m_{app} - i_n(0)(1 - \rho_1\psi_2)^{-2}/c_e$
m	Molecular mass extrapolated to zero concentration	$m = m_{app} + 2Bm^2c_e$
M_w	Weight-average molecular weight	$M = mN_A/q$
V	Hydrated volume of particle	$V = i_n(0) \bigg/ \int_0^\infty 2\pi s j_n(s) ds$
$\Delta\rho$	Electron density difference; solvent–solute	$\Delta\rho = \rho_1 - \rho_2 = \dfrac{\int_0^\infty 2\pi s j_n(s) ds}{c_e(1 - \rho_1\psi_2)} + c_e\rho_1(1 - \rho_1\psi_2)$
H	Degree of hydration in e- of bound H_2O per e- of particle	$H = \dfrac{\rho_1(1 - \rho_2\Psi_2)}{\Delta\rho}$

Symbol definitions:
c_e, concentration in amounts of e- of solute per e- of solution
ρ_1, electron density of solvent, 0.355 e-/Å^3 for water
ψ_2, electron partial specific volume of solute, $\psi_2 = \bar{v}/q$
$\bar{v}$, partial specific volume of the particle
q, number of electrons per gram of particle
B, second virial coefficient
N_A, Avogadro's number
ρ_2, mean electron density of hydrated solute

R_a = apparent radius of gyration of the macromolecule obtained by slit
collimation at a finite concentration of solute,

$\phi(s)$ = residual error between the model and the observed scattering.

The exponential term on the right-hand side of eq. (9.4) represents the
traditional Guinier approximation [3]. The radius of gyration R_a and $j_n(0)$
can be obtained by nonlinear regression of $j_n(s)$ vs. s data.

By comparing eq. (9.4) with the first entry in Table 3.8, we see that the
model for $j_n(s)$ vs. s is simply a *Gaussian peak* with $x_0 = s_0 = 0$, so that
$(x - x_0)^2 = (s - s_0)^2 = s^2$. The maximum of the Gaussian occurs at $s = 0$
(Figure 9.1), so that we see only the right-hand half of the peak.

In practice, proteins are often encountered in solutions in equilibrium
with their aggregates or as mixtures of proteins. Furthermore, asymmetry
associated with rod-shaped particles, for example, or spherically symmetric
particles with inhomogeneous regions of differing electron density may be
encountered. Any of these situations can lead to models with multiple
Gaussian components [4]. A model for such systems can be formulated by
summing terms similar to the exponential in eq. (9.4). Such a general model
is summarized in Table 9.1.

The important parameter, R_a, the apparent radius of gyration of the
macromolecule, can be obtained from nonlinear regression analysis of the
scattering data onto the model in Table 9.1. If the theoretical point-source
scattering function at zero angle $i_n(0)$ is obtained from these data, additional
parameters relating to the system can be found. Traditionally, $i_n(0)$ is obtained by deconvolution of the smeared infinite-slit data $j_n(s)$ and extrapo-
lating to zero angle. A computer program for this deconvolution has been
published [5]. Later in this chapter, we will present a method to obtain
$i_n(0)$ directly. The additional parameters that can be estimated from $i_n(0)$
are summarized in Table 9.2.

B. Applications

B.1. Fits with Two Contributing Components

Proteins that exist as monomers in solution can give a better fit to the
model in Table 9.1 for $p = 2$ that for $p = 1$. Possible reasons for this result
include inhomogeneity or asymmetry in the electron density of the particle.

Because this finding was initially rather unexpected, regression analysis
onto the models in Table 9.1 was investigated [4] for ribonuclease as a
standard, using scattering data computed from crystallographic coordinates
obtained from the Brookhaven Protein Data Bank. Scattered intensities
were calculated by means of the Debye equation using the crystallographic
coordinates and atomic radii. The resulting scattered intensities are shown
as crosses in Figure 9.2. These symbols are somewhat obscured by the

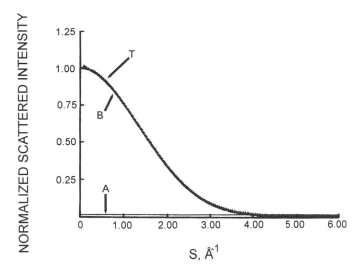

Figure 9.2 Scattering data for ribonuclease computed from its crystal structure (+) fit on the model in Table 9.1 with $p = 2$. Line T is best fit to model in Table 9.1 with $p = 2$; lines B and A are the individual Gaussian components computed from the regression parameters. (Reprinted from [4] with permission, copyright © 1988 Academic Press.)

regression line.) The fit to the model with $p = 1$ (not shown) has a relative root mean square error of 0.6% of the maximum and gives a radius of gyration of 14.28 Å. By contrast, the $p = 2$ fit, shown as a solid line, has a relative root mean square error of less than 0.08%, $R_{a,2} = 14.43 \pm 0.04$ Å and $R_{a,1} = 0.9 \pm 0.2$ Å. An extra sum of squares F test (see Section 3.C.1) showed that the $p = 2$ model was the best fit with >99% probability. This demonstrates clearly that a fit to a two-component model can result simply from the electron density characteristic of the protein's molecular structure.

B.2. SAXS Studies of Submicellar and Micellar Casein

B.2.a. Analysis of Scattering Data

The *caseins* occur in cow's milk as large colloidal complexes of protein and salts, commonly called *casein micelles*. Removal of calcium results in the dissociation of the micelle into smaller protein complexes called *submicelles*. The submicelles consist of four proteins, called $\alpha_{1,s}$-, $\alpha_{2,s}$-, β-, and κ-casein, in the ratios of 4:1:4:1. These proteins have average monomer molecular weights of 23,300 and are considered to have few specific secondary structural features. The isolated fractions exhibit varying degrees of self-association thought to be driven by hydrophobic interactions.

It has been hypothesized that, upon the addition of calcium, these hydrophobically stabilized, self-associated casein submicelles further self-associate via calcium–protein salt bridges. The resulting casein micelle has an average radius of 650 Å, as estimated by electron microscopy [6]. SAXS has been done on bovine casein to ascertain whether submicellar structures can be detected in the absence of calcium. The structure of the colloidal micelle in the presence of calcium was also investigated [4]. The results of this study will be examined in detail to illustrate the analysis of SAXS data. Scattering data for the buffer samples in this work [4] were fitted by the model in Table 9.1 with $p = 2$, and the fitted curves were subtracted as backgrounds from all protein scattering curves.

SAXS curves for submicellar and micellar casein are shown in Figure 9.3A. The different scattering magnitudes of these two samples are caused by the difference in protein concentration. Close inspection indicates that the shapes of these two curves are qualitatively different.

This difference for the two data sets is quite striking when the SAXS results are plotted in the traditional linearized (Guinier) form, as shown in Figure 9.3B. The submicelle plot is characterized by two linear regions, while the micelle data has three linear regions. However, an objective quantitative analysis of these data by the linear plot method is difficult for several reasons.

In the linearization of the SAXS models, there are difficulties in addition to changes in the error distribution of the dependent variable discussed in Section 2.A.3. When fitting a succession of straight lines to a slightly curving Guinier plot, as in Figure 9.3B, the determination of the location of a break in the curve is a matter for which objective critieria are rarely used. How do we decide where one straight line ends and the next one begins? In usual practice, a best guess at the answer to this question is used. Furthermore, the Guinier plot is an exponential approximation to a series expansion. It begins to deviate appreciably from a straight line at scattering angles beyond the

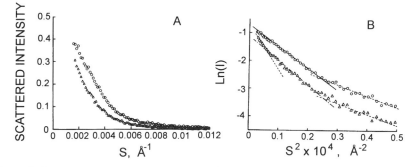

Figure 9.3 SAXS of submicellar casein (no Ca^{2+}) at 19.38 mg/mL (O) and micellar casein at 16.4 mg/mL (Δ): (A) absolute intensity; (B) Guinier plots. (Reprinted from [4] with permission, copyright © 1988 Academic Press.)

Guinier region, in the present case above $s = 0.0025$ Å. In other words, the exact model for the natural log of scattering intensity vs. s^2 is *not* a straight line!

As an alternative to Guinier plots, the use of nonlinear regression analysis onto the model in Table 9.1 provides objective and sensitive criteria for goodness of fit in the form of residual or deviation plots and the extra sum of squares F test (Section 3.C.1). This analysis of SAXS data affords a measure of statistical significance absent in a linearized plot, unless properly weighted linear regression and methods to find statistically best intersection points for multiple-line plots are employed. What is actually optimized in a linear Guinier plot is not the fit to the excess scattering intensity but to its logarithm. This tends to deemphasize the measurements at the smaller angles for which the precision is greatest. Also, the theoretical restriction to keep within the Guinier region is not a serious limitation with nonlinear fits. Here, the data at the higher angles make a much smaller contribution. Although the traditional linear plots may be useful for their heuristic and qualitative value, the analysis of SAXS data benefits greatly from nonlinear regression.

Data for submicellar casein gave best fits to the model in Table 9.1 with two Gaussian components ($p = 2$), as illustrated in Figure 9.4A. Also shown on the graph are the contributions of each of the two components as computed from the regression parameters. Goodness of fit is supported by the small standard deviation and the small standard error in each parameter (Table 9.3) and also by the random appearance of the residual or deviation plot (Figure 9.4B).

The standard deviation is well below the standard error of the measurement of ± 0.004. Note that, contrary to most examples in this book, the deviations have not been divided by SD for the residual plots in this section.

Table 9.3 Results of Nonlinear Regression of SAXS Data onto Models in Table 9.1[a]

Parameter	Casein submicelle	Casein micelle
$j_{n,1}(0)$	0.0326 ± 0.004	0.0283 ± 0.003
$R_{a,1}$	37.98 ± 0.004	34.7 ± 1.5
$j_{n,2}(0)$	0.439 ± 0.007	0.307 ± 0.011
$R_{a,2}$	81.7 ± 0.9	89.96 ± 0.01
$j_{n,3}(0)$		0.287 ± 0.046
$R_{a,3}$		204 ± 15
SD[b]	0.0022	0.0021

[a] Data from [4].
[b] Standard deviation of the regression analysis.

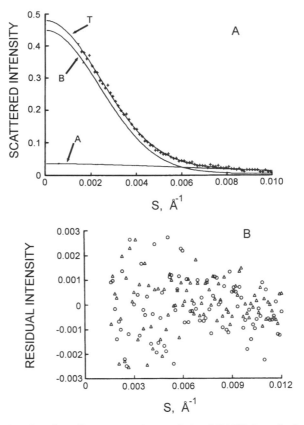

Figure 9.4 Results of nonlinear regression analysis of SAXS for submicelles of casein (19.4 mg mL^{-1}): (A) experimental scattering data ($+$); line T is best fit to model in Table 9.1 with $p = 2$; lines B and A are the individual Gaussian components computed from the regression parameters; (B) random residual plots from nonlinear regression onto $p = 2$ model of submicelle data in Figure 9.4A (O) and from regression onto $p = 3$ model for micelle data in Figure 9.6 (Δ). (Reprinted from [4] with permission, copyright © 1988 Academic Press.)

A different situation occurs when the submicellar data are fitted by a single component model ($p = 1$). In this case, the deviation plot (Figure 9.5) has a definite nonrandom pattern, indicating that an incorrect model was used for the fit. Moreover, the relative values of the deviations are about an order of magnitude larger than for the $p = 2$ model, reflecting the larger standard deviation of regression of the $p = 1$ fit. It is clear then, that even though the submicelle solution is known to contain a single particle, two Gaussian components are required to fit the data adequately. This result is similar to the finding from analysis of the scattering data computed from the crystal structure of ribonuclease.

The micellar casein data gave poor fits to the models in Table 9.1 with

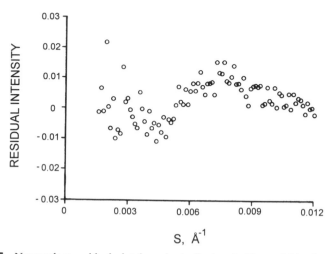

Figure 9.5 Nonrandom residual plot for submicelle data in Figure 9.4A after regression onto the single component model ($p = 1$) in Table 9.1. (Reprinted from [4] with permission, copyright © 1988 Academic Press.)

$p = 1$ or 2, but was fit successfully with the $p = 3$ model (Figure 9.6). Deviation plots were random for $p = 3$ and nonrandom for $p = 1$ or 2. The standard deviation of regression for $p = 3$ was well below the standard error of the measurements of ± 0.004. The deviation plot is in Figure 9.4B.

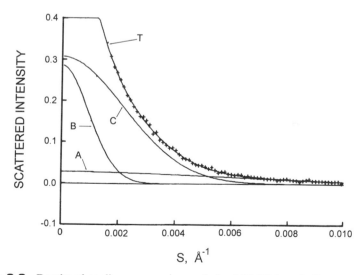

Figure 9.6 Results of nonlinear regression analysis of SAXS for micelles of casein (16.4 mg mL^{-1}): experimental scattering data ($+$); line T is best fit to model in Table 9.1 with $p = 3$; lines A, B, and C are the three individual Gaussian components computed from the regression parameters. (Reprinted from [4] with permission, copyright © 1988 Academic Press.)

All submicellar casein data examined were fit best by the $p = 2$ model (Table 9.1), while all micellar casein data were best fit by the $p = 3$ model. The values of the parameters were used to extrapolate the experimental data to zero scattering angle. The composite curves were then desmeared using the computer program developed by Lake [5]. This provides the deconvoluted intensity at zero angle, $i_n(0)$.

B.2.b. Interpretation of the Regression Analyses

The deconvoluted intensity at zero angle divided by the corresponding protein concentration yields a value proportional to the molecular weight of the particle (m_{app}, Table 9.2). This quantity, essentially $i_n(0)/c_e$, showed no concentration dependence under any of the conditions studied. Hence, extreme particle size polydispersity is unlikely. The particles are likely to have a single, narrow size distribution.

Other models accounting for the double-component ($p = 2$) character of the submicellar scattering data might be based on particle asymmetry (e.g., rods), or on a spherically symmetrical but inhomogeneous particle having regions of differing electron density. The former would not be in agreement with hydrodynamic, light scattering, and electron microscopic evidence, which indicates that submicellar casein exists as spherical particles [6].

Because the particles result from a hydrophobically driven self-association of monomer units, it has been considered most likely that they contain a hydrophobic inner core surrounded by a hydrophilic outer layer [6]. This arrangement would be the most stable on thermodynamic grounds. Such a structure would have different packing densities in regions where predominantly hydrophobic and hydrophilic amino acid side chains reside. This would give rise to two roughly concentric regions of different electron density [4] (Figure 9.7).

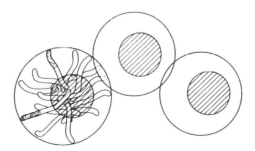

Figure 9.7 Schematic representation of casein submicelles within a micellar structure. Cross-hatched area represents core regions of higher electron density and higher concentration of hydrophobic amino acid side chains. The particle on the lower left shows a few representative casein monomer chains. (Reprinted from [4] with permission, copyright © 1988 Academic Press.)

Thus, the submicellar data were analyzed further by means of a model in which the particle has two regions of different electron density within the same scattering center. In this model, the scattered amplitudes rather than the intensities of the two regions must be added because of interference effects of the scattered radiation.

Luzzati et al. [1] developed equations for calculating the molecular and structural parameters of a particle having two regions of different electron density with the same electronic center of mass. These are applicable to smeared SAXS data containing two linear Guinier regions. The model is essentially identical in form to that in Table 9.1 with $p = 2$. They identified the parameters $j_{n,1}(0)$ and $R_{a,1}$ with the inner electron density region, and $j_{n,2}(0)$ and $R_{a,2}$ with the outer electron density region. They expressed the deconvoluted zero angle scattering intensity $i_n(0)$, by

$$[i_n(0)]_2 = 2\sqrt{\pi/3}\,[j_{n,1}(0)R_{a,1} + j_{n,2}(0)R_{a,2}]. \qquad (9.5)$$

The average radius of gyration, R_2, of the inhomogeneous particle (i.e., the whole submicelle) is obtained from

$$R_2^2 - 2\sqrt{\pi/3}\,[\bar{j}_{n,1}(0)R_{a,1}^3 + \bar{j}_{n,2}(0)R_{a,2}^3]^{-1}. \qquad (9.6)$$

In the next models, the subscripts C and L (for compact and loose) [1] are used to designate the higher and lower electron density regions, respectively, and the subscript 2 designates the entire particle composed of these two regions. The respective masses are

$$M_2 = M_C + M_L \qquad (9.7)$$
$$M_C = fM_2 \qquad (9.8)$$

where f is the fraction of electrons in the higher electron density region. The fraction f is easily evaluated from the relationship

$$[i(0)]_C = f^2[i(0)]_2 \qquad (9.9)$$

$[i(0)]_2 =$ deconvoluted scattering intensity at zero angle for the whole submicellar particle,
$[i(0)]_C =$ deconvoluted scattering intensity at zero angle for the higher electron density region.

In Table 9.3, $R_{a,1}$ can be identified with the radius of gyration of the denser region, R_C. Thus,

$$[i(0)]_C = 2\sqrt{\pi/3}\,R_{a,1}j_{n,1}(0). \qquad (9.10)$$

The radius of gyration of the low electron density region, R_L, can now be found from the expression

$$R_2^2 = fR_C^2 + (1 - f)R_L^2 \qquad (9.11)$$

R_2 is available from eq. (9.6), the fraction f can be obtained from eq. (9.9), and $R_C = R_{a,1}$ obtained from the regression analysis (Table 9.3).

Table 9.4   Molecular Parameters for Casein Aggregates

Parameter	Casein submicelle	Casein micelle
M		$882{,}000 \pm 28{,}000$
k_2		0.308 ± 0.005 (3.2:1)
M_2	$285{,}000 \pm 14{,}600$	$276{,}000 \pm 18{,}000$
k	0.212 ± 0.028	0.216 ± 0.003
M_C	$60{,}000 \pm 5{,}650$	$56{,}400 \pm 3{,}700$
M_L	$225{,}000 \pm 18{,}500$	$220{,}000 \pm 18{,}700$
$\Delta\rho(\text{e-}/\text{Å}^3)$		0.0081 ± 0.0004
$\Delta\rho_2(\text{e-}/\text{Å}^3)$	0.0099 ± 0.0004	0.0073 ± 0.0005
$\Delta\rho_C(\text{e-}/\text{Å}^3)$	0.0148 ± 0.0014	0.0128 ± 0.0007
$\Delta\rho_L(\text{e-}/\text{Å}^3)$	0.0091 ± 0.0003	0.0065 ± 0.0003
H ($g_{water}/g_{protein}$)		7.92 ± 0.42
H_2 ($g_{water}/g_{protein}$)	6.31 ± 0.30	8.98 ± 0.44
H_C ($g_{water}/g_{protein}$)	3.97 ± 0.48	4.70 ± 0.31
H_L ($g_{water}/g_{protein}$)	6.90 ± 0.64	9.95 ± 0.58

From these equations and Table 9.2, the molecular and structural parameters for casein under submicellar conditions were evaluated. No change in any molecular or structural parameter was observed with variation of the protein concentration. The averages of these results are presented in Table 9.4 for the molecular parameters and Table 9.5 for the structural parameters, subscripted as previously. These two tables illustrate the vast amount of quantitative information and the excellent precision available from analysis of SAXS on macromolecular systems with the aid of nonlinear regression.

A similar analysis was done on the SAXS results for the casein micelle solutions. The first two components of the model in Table 9.1 represented by the first four parameters in Table 9.3 reflect the contribution of the submicellar structure to the SAXS results. The third component, which has

Table 9.5   Structural Parameters from SAXS Data

Parameter	Casein submicelle	Casein micelle
V (Å^3)		$12.72 \pm 0.25 \times 10^6$
V_2 (Å^3)	$3.33 \pm 0.26 \times 10^6$	$4.44 \pm 0.16 \times 10^6$
V_C (Å^3)	$0.467 \pm 0.002 \times 10^6$	$0.529 \pm 0.003 \times 10^6$
V_L (Å^3)	$2.86 \pm 0.40 \times 10^6$	$3.91 \pm 0.03 \times 10^6$
R_2 (Å^3)	80.24 ± 0.39	90.57 ± 0.03
R_C (Å^3)	37.98 ± 0.01	39.62 ± 0.01
R_L (Å^3)	88.22 ± 0.82	100.2 ± 0.1

the highest radius of gyration ($R_{a,3}$ in Table 9.3), reflects the total number of submicellar particles within the cross-sectional scattering profile. Here, at zero angle, the intensity of the larger Gaussian contribution can be simply added to the intensity of submicellar contribution. A new parameter, f_2, the ratio of the mass of the submicelles to the total observed mass ascribable to a cross section, is defined as

$$f_2 = \frac{[i(0)]_2}{2\sqrt{\pi/3}\,R_{a,3}j_{n,3}(0) + [i(0)]_2}.$$

(9.12)

The inverse of f_2 is called the *packing number;* that is, the number of submicellar particles within the observed "cross-sectional mass."

The scattering data for casein micelles were analyzed using the preceding equations for all protein concentrations. Again, no variation of SAXS-derived parameters with protein concentration was observed. The average values with corresponding errors are presented in Tables 9.4 and 9.5. In these tables, subscript 2 denotes corresponding parameters for the submicellar particles when incorporated in an observed scattering volume of the micelle, and the unsubscripted parameters represent total cross-sectional features of the colloidal micelle particle.

The SAXS results on casein were further interpreted by using physical models for the submicelle and micelle [4]. Tertiary and quaternary structural differences between two genetic variants of bovine casein have also been studied by SAXS with nonlinear regression of the data [7] onto the models in Table 9.1. The preceding discussion illustrates that this approach yields a quite complete physical picture of the macromolecular species present in the samples.

References

1. V. Luzzati, J. Witz, and A. Nicolaieff, "Determination of Mass and Dimensions of Proteins in Solution by X-Ray Scattering Measured on an Absolute Scale," *J. Molec. Biol.* **3** (1961), pp. 367–378, "The Structure of Bovine Serum Albumin in Solution at pH 5.3 and 3.6; Study by Absolute Scattering of X-Rays," pp. 379–392.
2. K. E. Van Holde, *Physical Biochemistry.* Englewood Cliffs, N.J.: Prentice-Hall, 1971.
3. H. Pessen, T. F. Kumosinski, and S. N. Timasheff in C. H. W. Hirs and S. N. Timasheff (Eds.), *Methods in Enzymology,* vol. 27D, pp. 151–209. San Diego: Academic Press, 1973.
4. T. F. Kumosinski, H. Pessen, H. M. Farrell, and H. Brumberger, "Determination of the Quaternary Structural States of Bovine Casein by SAXS: Submicellar and Micellar Forms," *Arch. Biochem. Biophys.* **266** (1988), pp. 548–561.
5. J. A. Lake, *Acta Crystallogr.* **23** (1967), pp 191–194.
6. D. G. Schmidt, "Association of Caseins and Casein Micelle Structure," in P. F. Fox (Ed.), *Developments in Dairy Chemistry,* vol. 1. Essex, U.K.: Applied Sci. Publ., 1982, pp. 61–86.
7. H. Pessen, T. F. Kumosinski, H. M. Farrell, and H. Brumberger, "Tertiary and Quaternary Structural Differences between Two Genetic Variants of Bovine Casein by SAXS," *Arch. Biochem. Biophys.* **284** (1991), pp. 133–142.

Chapter 10

Ultracentrifugation of Macromolecules

A. Sedimentation

Ultracentrifugation involves sedimentation at high rotation speeds. This technique is routinely used for separating macromolecules and determining their molecular weights and for the analysis of mixtures of proteins [1]. In this chapter, we present models for sedimentation velocity and equilibrium experiments and discuss several application to protein biochemistry.

In ultracentrifugation, a homogeneous solution of macromolecules is placed into a sector-shaped cell like that shown in Figure 10.1. This cell is placed in a rotor that spins about the rotation axis at angular velocity ω. Centrifugal, buoyant, and frictional forces act upon the macromolecules in the cell. The interplay between these forces govern the displacement of the macromolecules during the ultracentrifugation [1].

In this chapter, we discuss of two types of ultracentrifuge experiments and applications of their use. Sedimentation velocity experiments can be used to obtain the sedimentation coefficient of a macromolecule, which depends on the molecular weight and the frictional coefficient. Sedimentation equilibrium experiments are used to obtain weight average molecular weights.

A.1. Sedimentation Velocity

At the start of an ultracentrifugation experiment, the macromolecules are distributed homogeneously in the test solution. When the rotor begins to spin, the molecules begin to be pushed toward the bottom of the sector-shaped cell (Figure 10.1). This creates a moving boundary that travels

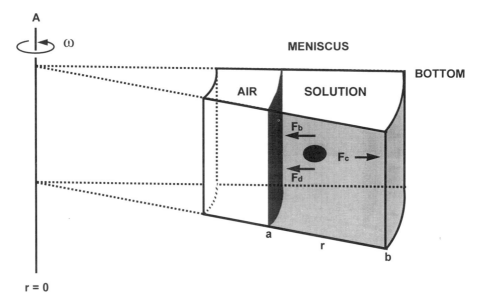

Figure 10.1 Diagram of a sector shaped cell (Adapted from Van Holde, *Physical Biochemistry,* © 1971, 99. Reprinted by permission of Prentice-Hall, Inc., Englewood Cliffs, N.J.)

toward the bottom of the cell. Above this boundary, there are no macromolecules, which all become concentrated below the boundary.

The rate of movement of the boundary depends upon the sedimentation coefficient s of the molecule. In the absence of diffusion, s is expressed as [1]

$$s = \frac{M_w(1 - \bar{v}\rho)}{N_A f} \tag{10.1}$$

where M_w is molecular weight, $\bar{v}$ is the partial specific volume of the molecule, ρ is the density of the solution, N_A is Avagadro's number, and f is the frictional coefficient. The velocity dr/dt of the moving boundary, where r is the distance from the center of rotation, is directly proportional to s:

$$\frac{dr}{dt} = r\omega^2 s \tag{10.2}$$

where r is the distance of the center of the boundary from the center of rotation and ω is the angular velocity. Integrating this equation gives

$$r(t) = r(t_0) \exp[\omega^2 s(t - t_0)]. \tag{10.3}$$

The use of eq. (10.3) requires an accurate determination of $r(t)$ at various times at a fixed angular velocity. The linearized version of eq. (10.3) is:

$$\ln\frac{r(t)}{r(t_0)} = \omega^2 s(t - t_0). \tag{10.4}$$

Thus, data in the form of $\ln[r(t)/r(t_0)]$ vs. $(t - t_0)$ could be subjected to linear regression to give $\omega^2 s$ as the slope. However, as discussed in Section 2.C.1, the error distributions in the independent variable will change accompanying linearization. If linear regression is to be used, it must be properly weighted. Again, the use of nonlinear regression directly onto eq. (10.3) avoids these problems.

The diffusion of macromolecules during sedimentation causes the moving boundary to spread with time [1, 2]. This spreading can be estimated by analysis of the shape of the boundary. Nonlinear regression can be employed for the estimation of spreading factors and $r(t)$ values during the sedimentation velocity experiment.

The model is the *integral* of the normal curve of error or Gaussian peak shape and can be computed from an integrated, series representation of the error function. [3, 4]. The spreading factor is the width at half-height of the integrated Gaussian peak, and $r(t)$ is the position of integrated peak. If s and D are constant throughout the boundary, then a graph of the square of the spreading factor (W^2) in cm^2 vs. time in seconds gives the apparent diffusion coefficient of the macromolecule in cm^2 s^{-1} as (slope)/2 of the linear regression line. The complete model for sedimentation velocity experiments assuming that absorbance is used as a detection method is given in Table 10.1.

Sedimentation velocity data for a solution of freshly purified bovine β_2-microglobulin shows excellent agreement with the model in Table 10.1 (Figure 10.2). Diffusion and sedimentation coefficients are obtained by analysis as described previously. M_W is obtained from s using eq. (10.1), after computing values of $\bar{v}$ and f from appropriate models [5–7]. These parameter values are generally corrected to 20°C in water. Analysis of a full set of $r(t)$ vs. $(t - t_0)$ data for bovine β_2-microglobulin[2] gave $s_{20,w} = 1.86$,

Table 10.1 Models for Analysis of Sedimentation Velocity Data for a Single Macromolecule

---

Assumptions: One macromolecule, no association or dissociation; detection by absorbance or refractive index
Regression equations:
  Position of boundary at t:
        series approximation to integral of Gaussian peak [2]
        $$X = \frac{r - r_0}{\sqrt{2}W} \qquad \eta = \frac{1}{1 + 0.47047X}$$
        $$A = h[1 - (0.3084284\eta + 0.0849713\eta^2 + 0.6627698\eta^3)\exp(-X^2)]$$
  Velocity of boundary: $r(t) = r(t_0)\exp[\omega^2 s(t - t_0)]$
Regression parameters: *Data:*
Position of boundary: h r_0 W A vs. r
Velocity of boundary: $r(t_0)$ s $r(t)$ vs. $(t - t_0)$
Special considerations: Better accuracy for A may be obtained by using more terms in the series [5]; D is obtained from linear plot of W^2 vs. t_2 as slope/2

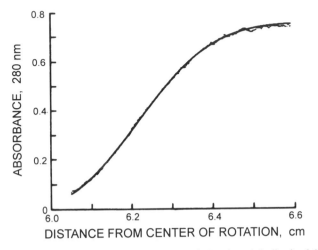

Figure 10.2 A sedimentation velocity pattern of β_2-microglobulin in 0.08 M NaCl, 0.02 M sodium phosphate, pH 5.0 at 25°C at 52,000 rpm for 96 min. in a 1.2 cm cell. Initial concentration <0.3 mg mL^{-1}. The points are experimental data, and the solid line is the best fit onto A vs. r model in Table 10.1. (Reprinted from [5] with permission.)

$D_{20,w} = 1.37 \times 10^{-6}$ cm^2 s^{-1}. Knowledge of these parameters allowed an estimation of $M = 11,900$.

Slow aggregation of proteins under conditions of storage is quite common and can lead to the presence of multiple boundaries in the sedimentation velocity experiment. These data can be modeled by summing the model for A vs. r in Table 10.1. This model is summarized in Table 10.2. An application of this model to bovine β_2-microglobulin [5] showed that this

Table 10.2 Model for Analysis of Sedimentation Velocity Data for Samples Containing Two Macromolecules

---

Assumptions: Two macromolecules, or association; detection by absorbance or refractive index

Regression equations:

Position of boundary at t:

      Series approximation to sum of two integrals of Gaussian peaks [2]

$$X_i = \frac{r - r_{o,i}}{\sqrt{2}W_i}; \quad \eta_i = \frac{1}{1 + 0.47047X_i}; \quad i = 1, 2$$

$$A = h_1[1 - (0.3084284\eta_1 + 0.0849713\eta_1^2 + 0.6627698\eta_1^3)\exp(-X_1^2)] +$$
$$h_2[1 - (0.3084284\eta_2 + 0.0849713\eta_2^2 + 0.6627698\eta_2^3)\exp(-X_2^2)]$$

Regression parameters: *Data*

  h_1 h_2 $r_{0,1}$ $r_{0,1}$ W_1 W_2 A vs. r

  Special considerations: D is obtained from plot of W^2 vs. t

---

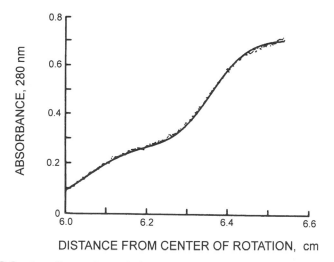

Figure 10.3 A sedimentation velocity pattern of β_2-microglobulin in 0.08 M NaCl, 0.02 M sodium phosphate after incubation for seven days at pH 5.0 at 25°C; 60,000 rpm for 40 min. in a 1.2 cm cell. The points are experimental data and the solid line is the best fit onto A vs. r model in Table 10.3. (Reprinted from [5] with permission.)

protein dimerized when stored at 25°C for a week (Figure 10.3). The two sigmoid-shaped components are clearly seen, and the results of analysis of the $r(t)$ vs. $(t - t_0)$ for each of them is given in Table 10.3. These data were used to show that the radius of the aggregate defined by the Stokes–Einstein equation (eq. (6.15)) was twice that of the monomer.

The models in Tables 10.1 and 10.2 are appropriate for ultracentrifuge experiments in which either absorbance or refractive index is used as the method of detecting the moving boundary. The absorbance is directly related to the concentration of the macromolecule. Another major detection method uses Schlieren optics, which measures the concentration gradient dC/dr. When this method is used, the models for the shape of the boundary

Table 10.3 Parameters for Irreversible Aggregation of Bovine β_2-Microglobulin from Sedimentation Velocity Experiments [6]

Species	$s_{20,w}$	$f/f_0{}^a$	Stokes radius,[b] Å
Monomer	1.21	1.56	17
Aggregate	5.10	1.48	34

[a] Normalized frictional coefficient.
[b] Computed as in [6].

are the derivatives of those presented in Tables 10.1 and 10.2. The model for the velocity of the boundary is the same irrespective of the detection method.

A.2. Sedimentation Equilibrium

When an ultracentrifuge experiment reaches equilibrium, there is no net transport of matter in the sector cell. Hence, the concentration boundary does not move. In the Yphantis method [6] of high-speed sedimentation equilibrium, the concentration of macromolecules is negligible at the meniscus in the cell and can be used as a point of reference. This is also called the *meniscus depletion method*. Analysis of sedimentation equilibrium data using this method will be discussed in this section.

The sedimentation equilibrium experiment for a single ideal solute can be described by

$$\int_{c_0}^{c} \frac{dc}{c} = \int_{r_a}^{r} \sigma_W r dr \qquad (10.5)$$

where the symbols are defined as follows:

$\sigma_W = M_W \omega^2 (1 - \bar{v}\rho)/RT$,
M_W = weight average molecular weight of the macromolecule,
$\bar{v}$ = partial specific volume of the macromolecule,
ρ = density of the solvent,
R = ideal gas constant,
T = temperature in Kelvins,
c_0 = concentration in mg mL^{-1} at the meniscus,
c = concentration at a distance r from the center of rotation,
r_a = distance of the meniscus from the center of rotation,
r = distance from the center of rotation,
ω = angular velocity of the rotor in radians s^{-1}.

Absorbance vs. distance data for sedimentation equilibrium can be analyzed by the integral of eq. (10.5). This model is summarized in Table 10.4.

Table 10.4 Model for Analysis of Sedimentation Equilibrium Data Obtained by the Yphantis Meniscus Depletion Method [6] for a Single Macromolecule

---

Assumptions: One macromolecule, no association or dissociation; detection by absorbance, A, which is proportional to c
Regression equations:
$$c = c_0 \exp[(\sigma_w / 2)(r^2 - r_a^2)]$$
Regression parameters: *Data:*
c_0 σ_w A vs. r
Special considerations: M_w is obtained from $\sigma_w = M_w \omega^2 (1 - \bar{v}\rho)/RT$; other symbols are defined in the text

---

We shall see that sedimentation equilibrium experiments can also be used to test for macromolecular aggregation. First, we show the results of an experiment on a monomeric macromolecule, holoRiboflavin-binding protein (Figure 10.4). This monomeric protein was found to have a M_W of 32,500 ± 1000 Daltons and to be free from aggregation or higher M_W impurities.

Proteins often associate reversibly to form dimers, trimers, and higher aggregates. Some typical association schemes follow:

$$M + M \rightleftharpoons M_2 \quad \text{(dimer)} \quad (10.6)$$
$$M + M_2 \rightleftharpoons M_3 \quad \text{(trimer)} \quad (10.7)$$
$$4\,M \rightleftharpoons M_4 \quad \text{(tetramer)} \quad (10.8)$$

A limiting model arises from the assumption that mainly monomer and n-mers exist in the solution:

$$n\,M \rightleftharpoons M_n \quad (n\text{-mer}). \quad (10.9)$$

The equilibrium constant for this association is given by

$$K = [M_n]/[M]^n. \quad (10.10)$$

This monomer/n-mer model is summarized in Table 10.5.

The model in Table 10.5 is illustrated for sedimentation equilibrium of bovine β_2-microglobulin (Figure 10.5). This protein forms a tetramer ($n = 4$) reversibly when the monomer concentration is above 0.4 mg mL^{-1}.

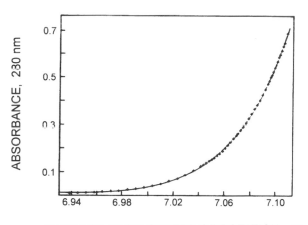

DISTANCE FROM CENTER OF ROTATION, cm

Figure 10.4 Sedimentation equilibrium data for holoRiboflavin binding protein, pH 7.0 in 0.1 M NaCl, 0.03 M sodium phosphate at 25°C; 8,000 rpm in a 1.2 cm cell. The points are experimental data, and the solid line is the best fit of the model in Table 10.4. (Reprinted from [7] with permission.)

Table 10.5 Model for Analysis of Sedimentation Equilibrium Data Obtained by the Yphantis Meniscus Depletion Method [6] for Monomer/n-Mer Equilibrium

---

Assumptions: Detection by absorbance A; $nM \rightleftharpoons M_n$
Regression equations:
$$c = c_0 \exp[(\sigma_w/2)(r^2 - r_a^2)] + nKc_0^n \exp[(n\sigma_w/2)(r^2 - r_a^2)]$$
Regression parameters: *Data:*
   c_0 σ_w K A vs. r
Special considerations: M_w is obtained from $\sigma_w = M_w\omega^2(1 - \bar{v}\rho)/RT$; other symbols are defined in the text. If the model in Table 10.4 gives a poor fit, do a series of regression with integral n values beginning with $n = 2$. Choose the model with the smallest standard deviation of regression as the best fit

---

The models in Table 10.4 and 10.5 are appropriate for ultracentrifuge experiments in which absorbance is used as the method of detecting the boundary. When Schlieren optics are used, the models for the shape of the boundary are the derivatives of those presented in Table 10.4 and 10.5. Alternatively, if the Schlieren data are integrated, the models in Table 10.4 and 10.5 would be applicable.

The model in Table 10.4 assumes that the sedimentation experiments were done on ideal systems. For nonideal systems, virial coefficients must be considered. These and other aspects of analysis of data from sedimentation equilibrium experiments have been discussed in detail [4].

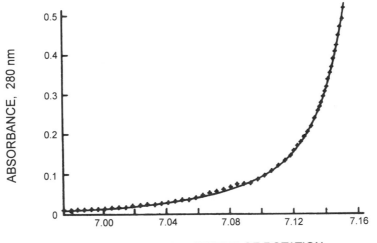

DISTANCE FROM CENTER OF ROTATION, cm

Figure 10.5 Sedimentation equilibrium data for β_2-microglobulin in 0.08 M NaCl, 0.02 M sodium phosphate, pH 5.0 at 25°C at 40,000 rpm in a 0.6 cm cell. Initial concentration >0.4 mg mL^{-1}. The points are experimental data, and the solid line is the best fit ($n = 4$) onto model in Table 10.5. (Reprinted from [5] with permission.)

References

1. K. E. Van Holde, *Physical Biochemistry.* Englewood Cliffs, N.J.: Prentice-Hall, 1971.
2. H. K. Schachman, *Ultracentrifugation in Biochemistry,* pp. 128–133. New York: Academic Press, 1959.
3. C. Hastings, J. T. Hayward, and J. P. Wong, *Approximations for Digital Computers.* Princeton, NJ: Princeton University Press, 1955.
4. M. L. Johnson and M. Straume, "Comments on the Analysis of Sedimentation Equilibrium Experiments," in T. M. Schuster and T. M. Laue (Eds.), *Modern Analytical Ultracentrifugation,* pp. 37–65. Boston: Birkhauser, 1994.
5. T. F. Kumosinski, E. M. Brown, and M. L. Groves, "Solution Physicochemical Properties of Bovine-β2- Microglobulin," *J. Biol. Chem.* **21** (1981), pp. 10949–10953.
6. D. A. Yphantis, "Equilibrium Ultracentrifugation of Dilute Solutions," *Biochemistry* **3** (1964), pp. 297–317.
7. T. F. Kumosinski, H. Pessen, and H. M. Farrell, "Structure and Mechanism of Action of Riboflavin-Binding Protein," *Arch. Biochem. Biophys.* **214** (1982), pp. 714–725.

Chapter 11

Voltammetric Methods

A. General Characteristics of Voltammetry

Voltammetric methods are a collection of electroanalytical techniques in which the potential of an electrolytic cell is varied toward negative or positive values and the current is measured [1, 2]. The output of the experiment is the current vs. the potential applied to the cell. Typical cells consist of reference and counter electrodes and a working electrode. The reference electrode is designed to hold a constant potential, and a potentiostatic circuit is used so that a variation in the potential applied to the cell changes the oxidizing or reducing power of the working electrode. One example of voltammetry was discussed in Section 3.B.1.

Different types of voltammetry are characterized by the different types of potential waveforms applied to the cell. For example, if the applied potential is varied linearly and the current measured continuously, the method is called *linear sweep voltammetry*. If a series of pulses of increasing heights is applied to the cell and the current measured at the end of each pulse, the technique is called *normal pulse voltammetry*. There are a wide variety of voltammetric methods [1]. Characteristics of some of those discussed in this chapter are summarized in Table 11.1.

A typical input potential waveform for a steady state voltammetric method such as rotating disk voltammetry is shown in Figure 11.1. The slope of this line in mV s^{-1} is called the *scan rate*. For a reversible, fast electron transfer reaction from reductant R to the electrode to yield oxidant O:

$$R \rightleftharpoons O + e^- \qquad E^{o\prime}. \qquad (11.1)$$

The current vs. potential ouput has the shape in Figure 11.2. $E^{o\prime}$, called the *formal potential,* is the apparent standard potential of the redox couple O/R measured in the solution of interest. Figure 11.2 points out the position

Table 11.1 Characteristics of Voltammetric Methods

Type of voltammetry	Potential input to the cell	Current measurement	Comments
Linear sweep	Linear variation, one direction	Continuous	Stationary electrode
Rotating disk	Linear variation, one direction	Continuous	Rotating electrode
Normal pulse	Increasing pulses	End of each pulse	Decreased charging current
Polarography[a]	Linear variation, one direction	Continuous	Dropping mercury electrode
Cyclic	Linear variation, one direction; then reverses scan	Continuous	Stationary electrode
Square wave	Forward–reverse square pulse on top of a staircase	End of each pulse	Best signal to noise and resolution
Differential pulse	Forward–reverse square pulse on top of a ramp	End of each pulse	Good signal to noise and resolution
Ultramicroelectrode[b]	Any of the preceding	Either of preceding	Electrode has dimension in μm range

[a] Any voltammetric method using a dropping mercury electrode as the working electrode is a type of polarography; for example, normal pulse polarography or square wave polarography.

[b] Discussed in detail in Section B.2.

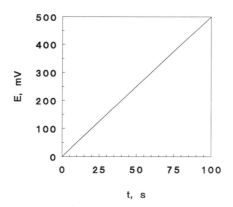

Figure 11.1 Typical potential ramp applied to an electrolytic cell for techniques such as rotating disk voltammetry and linear sweep voltammetry.

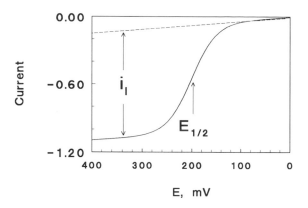

Figure 11.2 Current vs. potential output for a reversible oxidation in a *steady state* voltametric method.

of the half-wave potential $E_{1/2}$ as well as the definition of the so-called limiting current i_l, which is proportional to the amount of R dissolved in solution.

Note that the convention used for plotting voltammograms in this book has cathodic current as positive and anodic current as negative, and a potential scale that is negative to the right and positive to left [1, 2]. This convention is

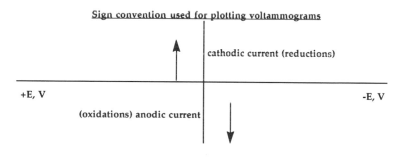

Figure 11.3 shows the input potential waveform for normal pulse voltammetry. The current is measured at the end of each pulse of applied potential. The shape of a reversible normal pulse current vs. potential ouput is the same as in Figure 11.2, except that the data are discrete points from one current measurement at the end of each pulse. Figure 11.4 shows the typical output from a reversible normal pulse voltammogram (NPV). In general, all pulse volammetric methods have a discrete digital output.

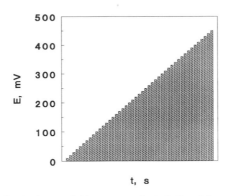

Figure 11.3 Typical pulsed potential input to an electrolytic cell for normal pulse voltammetry (NPV).

B. Steady State Voltammetry

B.1. Reversible Electron Transfer

The term *steady state voltammetry* is used in this book to denote any voltammetric method in which a steady state is achieved in some range of applied potential between the rate of electrolysis and some other feature of the experiment, such as the rate of diffusion or the rate of a limiting chemical reaction. This condition is signaled by the appearance of a plateau or limiting current in the voltammogram (cf. Figures 11.2 and 11.4). Steady state techniques include rotating disk voltammetry, dc polarography, normal pulse voltammetry, and slow scan ultramicroelectrode voltammetry.

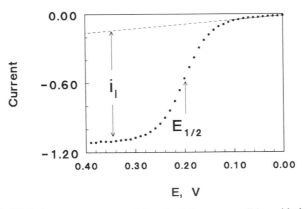

Figure 11.4 Digital current vs. potential output for a reversible oxidation in normal pulse voltammetry.

A somewhat general approach to analyzing steady state voltammograms can be used. This is because the regression equations for a reversible electrode reaction (cf. eq. (11.1)) have the same form for each steady state technique. They differ only in the expressions for the limiting current.

Conventional disk-shaped electrodes of radii larger than 0.1 mm develop faradaic currents in analytical electrochemical cells that are described to a good approximation by using a semi-infinite linear diffusion model. Expressions for their limiting currents in steady state voltammetry are available in standard electrochemistry textbooks [1, 2]. For reversible electrode reactions, the limiting currents are relatively independent of scan rate. Voltammograms obtained with microelectrodes, on the other hand, may change shape depending on the scan rate used. These electrodes are discussed in more detail in the next section.

We previously encountered the general model for reversible electron transfer in steady state voltammetry in Section 3.B.1. It is summarized in Table 11.2. This table introduces a second background option in addition to the linear one in Section 3.B.1. The linear ($b_4E + b_5$) term works well if the background is relatively flat, such as in Figures 11.2 and 11.4. However, a number of background processes on electrodes limit the available potential window, such as electrolysis of hydrogen ions, solvent, electrolyte ions, or the electrode itself, and can contribute to a large exponential background current.

The exponential term (Table 11.2) describes the rising signal caused by the irreversible reaction responsible for the background current at the potential limit of the working electrode. In such cases, the model is

$$i = i_l/(1 + \theta) + b_4[\exp(b_5E)]. \tag{11.2}$$

A voltammogram overlapped with an exponential background is illustrated for an oxidation wave in Figure 11.5. The resolved voltammogram was computed from the regression parameters after nonlinear regression analysis of the overlapped voltammetric data onto eq. (11.2).

Table 11.2 Model for Reversible Electron Transfer in Steady State Voltammetry

---

Assumptions: Reversible electron transfer, linear diffusion
Regression equation (*reductions*):
$$i = i_l/(1 + \theta) + \text{background}; \qquad \theta = \exp[(E - E_j^{o\prime})(F/RT)],$$
R = gas constant, F = Faraday's constant, and T = temperature in Kelvins

Regression parameters:			*Data:*
i_l	$E^{o\prime}$	(F/RT)	i vs. E

Special instructions: Linear background = $b_4E + b_5$
   Exponential background current = $b_4[\exp(b_5E)]$
   Use $\theta = \exp[-(E - E_j^{o\prime})(F/RT)_j]$ for oxidations

---

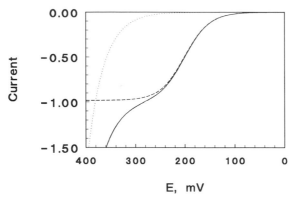

Figure 11.5   Steady state voltammogram for an oxidation wave (———) severely overlapped with a large increasing anodic current showing the resolved voltammogram computed from nonlinear regression parameters (--------) using model in Table 11.2 with exponential background and the separate exponentially increasing background (·········).

A similar model can be used when two or more voltammograms are overlapped with one and other (Figure 11.6). They can be separated by using nonlinear regression onto the model in Table 11.3, which is simply the sum of two reversible waves. The fraction (f) of component 1 is referred to the *total amount of analyte* in the mixture. Regression onto this model was used successfully to resolve overlapped dc polarograms for Pb(II) and Tl(I), which differ in half-wave potential by only about 50 mV [3].

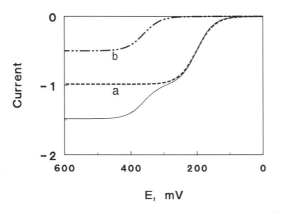

Figure 11.6   Severely overlapped steady state voltammograms representing two reversible oxidations (———) showing the two resolved voltammograms computed from the regression parameters using model in Table 11.3: (a) first component and (b) second component.

Table 11.3 Model for Two Overlapped Reversible Waves in Steady State Voltammetry

---

Assumptions: Two reversible electron transfers, linear diffusion
Regression equation (*reductions*):

$$i = i_1(f/(1 + \theta_1) + (1 - f)/(1 + \theta_2); \qquad \theta_j = \exp[E - E_j^{o'})(F/RT)_j], \qquad j = 1 \text{ or } 2;$$

R = gas constant, F = Faraday's constant, T = temperature in Kelvins, f = fraction of component 1 in the solution

Regression parameters: *Data:*

  f i_1 $E_1^{o'}$ $E_2^{o'}$ $(F/RT)_1$ $(F/RT)_2$ i vs. E

Special instructions: Linear background = $b_7E + b_8$

  Exponential background current = $b_7[\exp(b_8E)]$

  Use $\theta_j = \exp[-(E - E_j^{o'})(F/RT)_j]$ for oxidations

---

B.2. Reversible Electron Transfer at Ultramicroelectrodes in Highly Resistive Media

Electrodes with one dimension in the micrometer range have become popular tools in electrochemistry [4]. Platinum, gold, or carbon disks of radii 1–25 μm imbedded in glass or polymer insulating materials are typical of this class. Because very small currents are generated at electrodes of such very small size, they can be used in highly resistive media and at very high rates of scanning. Studies in organic solvents without electrolyte become possible. Efficient diffusion and small capacitance at these tiny electrodes leads to signal to noise that is often better than for larger solid electrodes of the same material. The possibility for scan rates on the order of 1 million V s^{-1} provides the ability to study very fast electrochemical and chemical reactions.

Microdisk electrodes of radii 1–25 μm give a steady state voltammetric response at low scan rates (e.g., <25 mV s^{-1}) when the electrode reaction is controlled by diffusion. The model for diffusion under such experimental conditions is essentially spherical, resulting in a steady state sigmoid shape of the current vs. potential curve. As the scan rate in such a system increases, the shape of the curve begins to show a peak, until at very large scan rates the response is peak shaped as in cyclic voltammetry. The curve shape changes because, as the scan rate increases, the time scale of the experiment becomes smaller and the spherical diffusion field collapses to the semi-infinite planar model. A steady state is no longer achieved, and depletion of electroactive material very near the electrode leads to a decrease in current after the peak [4].

The regression model for the sigmoid-shaped reversible response at low scan rates is the same as for any reversible steady state voltammogram (Table 11.2). Excellent fits of this model to microelectrode voltammograms representing standard reversible electrochemical reactions have been reported [5, 6].

Even though microelectrodes can be used in resistive fluids without adding electrolyte, the resulting voltammograms are somewhat broadened by the ohmic or iR_u drop of the cell, where $i =$ current and $R_u =$ uncompensated cell resistance. Resistance is large in solvents and oil-based media not containing sufficient electrolyte and is difficult to compensate for electronically. Therefore, the resulting ohmic broadening can be significant.

The ohmic drop of a cell is difficult to distinguish from slow electrode kinetics. However, in a microelectrode cell, a known reversible redox couple may be used to obtain R_u. Although the dependence of ohmic drop on potential in a cell containing a microdisk electrode may be complex, a simple approximate expression [5, 6] is effective when the ohmic drop is not too large:

$$E = V + iR_u \qquad (11.3)$$

where E is the actual potential at the microelectrode, and V is the applied voltage.

Equation (11.3) is combined with the model in Table 11.2 to yield the model in Table 11.4 for microelectrode voltammetry in a resistive medium. Note the special pair of logical statements that must be included in the subroutine that computes the current to avoid computer overflows or underflows. An overflow will occur when the result of a calculation exceeds the largest number the computer can store, thus triggering an error message and program termination. Analogously, some computers do not automatically assign a value of 0 to results that are too small for them to represent. This results in an underflow error message.

The model in Table 11.4 was tested with data obtained at scan rates of $\leq 10\ \text{mV s}^{-1}$ for the reversible oxidation of ferrocene in acetonitrile without added electrolyte at a carbon disk electrode of 6 μm radius [6]. The model in Table 11.4 fixes RT/F at its theoretical value and involves five parameters,

Table 11.4 Model for Reversible Ultramicroelectrode Voltammetry in a Resistive Medium

---

Assumptions: Reversible electron transfer, resistive media
Regression equation (reductions):
$$i = i_l/(1 + \theta) + b_4E + b_5; \qquad \theta = \exp[(V + iR_u - E^{o\prime})(F/RT)],$$
$R =$ gas constant, $F =$ Faraday's constant, and $T =$ temperature in Kelvins

Regression parameters:					*Data:*
R_u	i_l	$E^{o\prime}$	b_4	b_5	i vs. E

Special instructions: To avoid possible overflow or underflow, include the following in subroutine for computing i:
  IF $(E - E^{o\prime}) \leq -10$ THEN $i = i_l + b_4E + b_5$
  IF $(E - E^{o\prime}) \geq -10$ THEN $i = b_4E + b_5$
  use $\theta = \exp[-V + iR_u - E^{o\prime})(F/RT)]$ for oxidations

---

including a linear background. In testing the model, a modified steepest descent algorithm gave better convergence than the Marquardt–Levenberg algorithm because matrix inversion errors were generated with the latter. Analysis of 10 data sets with a program employing the steepest descent algorithm gave an average R_u of $3.47 \pm 0.20 \times 10^6$ ohms. Graphical representation of a typical data set shows that the points fit the regression line quite well (Figure 11.7). A resistance-free voltammogram computed from the results of the regression analysis is also presented.

The model in Table 11.4 has also been used to find the resistance of a *microemulsion* consisting of tiny surfactant-coated water droplets suspended in oil [7]. This medium had an R_u approximately 40-fold larger than acetonitrile in the preceding example. Knowledge of R_u allowed computation of resistance-free voltammograms for catalysts dissolved in the water droplets of the microemulsion and estimation of their electrochemical parameters.

B.3. Electron Transfer Rate Constants

B.3.a. Rotating Disk and Ultramicroelectrode Voltammetry

If the speed of the electrochemical experiment is so fast that the electron transfer step does not remain at equilibrium, the shapes of steady state voltammograms are controlled by kinetics as well as diffusion. A simple electrode reaction with slow kinetics with respect to the measurement time, for example,

$$R \rightleftharpoons O \mid e^-$$

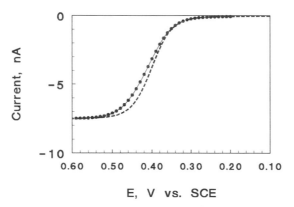

Figure 11.7 Steady state voltammogram at 7 mV s^{-1} of 1 mM ferrocene at a carbon microdisk electrode of 6 μm radius in acetonitrile with no added supporting electrolyte. The solid line shows experimental data with points computed from regression analysis using the model in Table 11.4. The dashed line was computed after mathematical removal of iR_u from the data. (Adapted with permission from [6], copyright by VCH.)

can be characterized by the apparent standard heterogeneous electron transfer rate constant, $k^{o\prime}$ in cm s^{-1}, and α, the electrochemical transfer coefficient [1, 2].

The electrochemical transfer coefficient is a number between 0 and 1, which is related to the symmetry of the energy barrier to electron transfer. A symmetric energy barrier is characterized by $\alpha = 0.5$. The apparent standard heterogeneous electron transfer rate constant ($k^{o\prime}$) is characteristic of the rate at which electrons are exchanged between electrodes and electroactive compounds at the formal potential $E^{o\prime}$. The value of $k^{o\prime}$ is called an *apparent* value because it is highly dependent on experimental conditions such as type of electrolyte, electrode materials, method of surface preparation, electrode contaminants, and surface funtionality. Values of $k^{o\prime} > 10$ cm s^{-1} are considered to be very large and are characteristic of electrochemically reversible systems, except for exceeding short experimental time frames. Values smaller than about 10^{-6} cm s^{-1} denote very slow electron transfer reactions.

A common model can be used for rotating disc voltammetry (RDV) and ultramicroelectrode steady state voltammetry when the electrode reactions are controlled by a mixture of electron-transfer kinetics and diffusion. The model is summarized in Table 11.5. The models for the two techniques differ in the definitions of the mass transfer coefficients and the limiting currents. Limiting currents can be used to estimate the diffusion coefficients needed for computing $k^{o\prime}$.

Table 11.5 Models for Obtaining Electron Transfer Rate Constants from RDV and Ultramicroelectrode Voltammetry

---

Assumptions: Rotation disk or microelectrode voltammogram controlled by electrode kinetics and diffusion; α does not depend on E

Regression equation:

$$i = i_l/(1 + \theta + k^\prime \theta^\prime) + b_4 E + b_5;$$
$$\theta = \exp[(E - E^{o\prime})(F/RT)]; \qquad k^\prime = k_m/k^{o\prime} \text{ (reductions)}$$
$$\theta^\prime = \exp[(E - E^{o\prime})(F/RT)\alpha] \text{ (reductions)}$$
$$\theta = \exp[-(E - E^{o\prime})(F/RT)]; \qquad k^\prime = k_m/k^{o\prime} \text{ (oxidations)}$$
$$\theta^\prime = \exp[-(E - E^{o\prime})(F/RT)\alpha] \text{ (oxidations)}$$

R = gas constant, F = Faraday's constant, T = temperature in Kelvins; k_m = mass transfer coefficient

Regression parameters: *Data:*

  i_l $E^{o\prime}$ α $k^\prime$ b_4 b_5 i vs. E

Special instructions: Keep (F/RT) fixed;

definitions of k_m depends on method: RDV $k_m = D_0/\delta_0$; $k^{o\prime} = D_0/\delta_0 k^\prime$ and
$$i_l = 0.62 n F A D_0^{2/3} \omega^{1/2} \nu^{-1/6} C; \qquad \delta_0 = 1.61 D_0^{1/3} \omega^{1/2} \nu^{-1/6}$$

steady state disk microelectrode; approximately: $k^{o\prime} = 1.226 \dfrac{D_0}{k^\prime r}$ and $i_l = 4 n F D_0 C r$

r = microdisk radius; D_0 = diffusion coefficient of reactant; δ_0 = diffusion layer thickness; ω = angular velocity; ν = kinematic viscosity; obtain D_o values from limiting currents [1, 2]

The model in eq. (11.5) has been used for RDV on conventional solid disk electrodes for a number of redox couples with intermediate values of $k^{o'}$. Results have been summarized briefly [5]. A good signal to noise ratio is essential for the success of the method, because background currents on solid electrodes can be large and have irregular shapes.

The model's use is now illustrated in more detail for oxidation of ferrocyanide at a carbon microdisk electrode of radius 6 μm. A graphical representation of regression analysis onto the model in Table 11.5 show a good fit to the data (Figure 11.8a). The deviation plot was random (Figure 11.8b). Although $k^{o'}$ will depend on the method of electrode pretreatment, results agree well with other determinations of $k^{o'}$ for the ferri/ferrocyanide redox couple on similarly pretreated carbon electrodes (Table 11.6). The upper limit of $k^{o'}$ estimations with this method is about 1 cm s^{-1} for electrodes with radii of 1–10 μm.

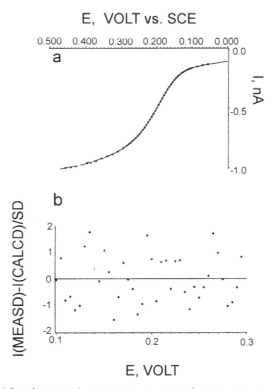

Figure 11.8 (a) Steady state voltammogram at 10 mV s^{-1} of 0.5 mM potassium ferrocyanide at a carbon microdisk electrode of 6 μm radius in 0.5 M KNO$_3$. The solid line shows experimental data with points computed from regression analysis using the model in Table 11.5. (b) The residual plot from the analysis in (a). (Reproduced with permission from [6], copyright by VCH.)

Table 11.6 Results of Regression Analysis for Ferri/Ferrocyanide Couple Compared to Published Data for Comparably Pretreated Surfaces

Method	$10^6 D$, cm^{2}s^{-1}	$10^2 k^{o\prime}$, cm s^{-1}	$E^{o\prime}$, V/SCE	Electrode/ electrolyte	Reference
c disk[a] Microelectrode	7.0	2.3	0.189 ± 0.004	6 μm[a]/ 0.5 M KNO$_3$	[6]
CV	7.6[c]			Pt, 1 M KCl	[8]
CV		3	0.21	GC[d], 2M KCl	[9]
NPV on GC[b] electrode	6.9 ± 0.5	3.0 ± 0.9	0.218 ± 0.007	0.5 M KNO$_3$	[10]

[a] Electrochemically pretreated carbon microdisk, steady state conditions.

[b] Conventionally polished glassy carbon using 1.1 mM potassium ferricyanide in 0.5 M KNO$_3$ (extensively polished and cleaned glassy carbon may give values 3 to 4 times larger than those quoted for ferrocyanide, see [11].

[c] M KCl has viscosity within 2% of 0.5 M KNO$_3$, and should give comparable D values.

[d] Conventionally polished and electrochemically pretreated.

B.3.b. Normal Pulse Voltammetry

This technique has advantages in the estimation of electron transfer rate constants compared to other voltammetric methods. The signal to noise ratio of the pulse techniques is generally better than in methods in which current is measured continuously. NPV is particularly suitable for regression analysis because it has a closed form model for voltammograms under diffusion–kinetic control.

In normal pulse voltammetry, a series of periodic pulses of linearly increasing or decreasing potential are applied to the electrochemical cell (cf., Figure 11.3). All pulses begin at the same base potential, at which there is no electrolysis at the working electrode. At the end of each pulse, the potential returns to the base value and a time lag precedes the next pulse. Current measurements are made at the end of each pulse, allowing the unwanted charging current to decay relative to the analytical faradaic current. A resulting increase in faradaic to charging current ratio in the current vs. potential curve, relative to methods such as cyclic voltammetry that measure current continuously, means that the signal to noise ratio is often larger and background currents are more well behaved and easier to model or subtract.

The model for NPV has been described by Osteryoung, O'Dea, and Go [12, 13]. It assumes that the electrode reaction under consideration is

$$O + n\,e^- \rightleftharpoons R$$

or its reverse. The model is summarized in Table 11.7.

The term $x \exp(x^2) \text{erfc}(x)$ in Table 11.7 can be computed by Oldham's approximation [14] to an accuracy of better than ±0.2%. Box 11.1 gives

Table 11.7 Model for Obtaining Electron Transfer Rate Constants from Normal Pulse Voltammetry

---

Assumptions: Voltammogram controlled by electrode kinetics and linear diffusion; α does not depend on E; D values of O and R are equal

Regression equations:

$$i(t) = i_d \pi^{1/2} x \exp(x^2) \, \text{erfc}(x)/(1 + \theta) + b_5 E + b_6$$
$$\theta = \exp[nF/RT](E - E^{\circ\prime}) \text{ (reductions)}$$
$$i_d = nFAD^{1/2}C_0/(\pi^{1/2}t_p^{1/2})$$
$$x = \kappa (1 + \theta)\theta^{-\alpha}t_p^{1/2}$$
$$\kappa = k^{\circ\prime}/D^{1/2}$$

R = gas constant, F = Faraday's constant, T = temperature in Kelvins; i_d = limiting current, C_0 = concentration of electroactive species, α = electrochemical transfer coefficient, t_p = pulsewidth, D = diffusion coefficient, and A = electrode area

Regression parameters: *Data:*

  i_d $E^{\circ\prime}$ α κ b_5 b_6 i vs. E

Special instructions: Keep (F/RT) fixed; use optimum pulsewidth range so that voltammogram is not reversible or totally irreversible; obtain D_0 values from limiting currents [1, 2]; $x \exp(x^2)\, \text{erfc}(x)$ is computed by Oldham's approximation [14]; use $\theta = \exp[-(nF/RT)(E - E^{\circ\prime})]$ for oxidations

---

Subroutine 11.1. BASIC subroutine for analysis of quasireversible NPV data with background based on Model in Table 11.7

```
10 THETA = EXP(38.92*(A(2,I%)-B(0)))
12 IF B(2) <0 THEN B(2) = 1
15 IF B(3)>1.02 THEN B(3) = .5
17 IF B(3)<0 THEN B(3) = .05
18 IF B(4)<0 THEN B(4) =.02
20 XXX = B(2)*(1 + THETA)*THETA^(-B(3))*SQR(TPULSE)
30 PXX = .8577*(1-(.024*XXX^2)*(1-.8577*XXX/3.1415927#^2))
40 PHI = 1 - (1-2/3.1215927#)*EXP(-XXX*PXX)
50 FXX = 2/(1 + (1 + PHI*2*XXX^(-2))^.5)
60 YC = D(1)*THO(/(1 | THIETA) - D(4)*(A(2,I%)) + D(5)
```

Documentation:

parameters: B(0) = $E^{\circ\prime}$; B(1) = i_d; B(2) = κ; B(3) = α; B(4) = background slope;

     B(5) = background intercept.

line 10 - computes θ

lines 12-18 - sets limits on parameters to avoid unproductive parameter space.

line 20 - computes x

line 30 - 50 computes $\pi^{1/2}$ x $\exp(x^2)\text{erfc}(x)$ by Oldham's approximation

line 60 - computes NPV current at potential E

Box 11.1 BASIC subroutine for use in nonlinear regression analysis of quasireversible NVP data with a background based on the model in Table 11.7.

the BASIC code for computing the NPV current–potential curve employing Oldham's approximation (Table 11.7) within a nonlinear regression program.

The subroutine in Box 11.1 was used in the analysis of data for the ferri/ferrocyanide redox couple on glassy carbon electrodes in 0.5 M KNO_3, which was discussed earlier for carbon microdisk electrodes. Table 11.6 shows that the results from the NPV analysis [10] agree well with those of microelectrode voltammetry and with other studies using similarly activated glassy carbon electrodes.

B.3.c. Errors in κ

If we are interested mainly in $k^{o\prime}$, it is essential to obtain the kinetic parameter κ with good accuracy. We must consider the influence of pulsewidth on the error in rate constant $k^{o\prime}$, which is contained in κ in the model in Table 11.7. This was done by analyzing theoretical data with normally distributed noise of $\pm0.5\%$ of the limiting current [10]. Because the true parameters are known for these data sets, their errors are the differences of the true values from those computed from nonlinear regression. Results show that the correct choice of pulse width in NPV is critical for accurate estimations of $k^{o\prime}$.

As pulsewidth is increased above about 10 ms for data with $\kappa = 12$, the shape of the current–potential curve eventually becomes reversible and the kinetic information in it decreases. For data with 1–2 ms pulsewidths, errors in κ from the regression analysis were about $\pm1\%$. In contrast, analysis of data with a pulse width of 30 ms gave a 20% error in κ. Hence, data for $\kappa = 12$ and $t_p \geq 30$ ms gave excellent fits to the reversible model in Table 11.2.

Data representing $\kappa = 12$ and three values of α show an optimum range of pulsewidths for obtaining κ with minimum errors (Figure 11.9). This optimum range does not seem to depend critically on α. Errors in κ for $\alpha = 0.35$ and $\alpha = 0.5$ fall on nearly the same line. However, the $\alpha = 0.65$ data showed an extended optimum range in the small pulsewidth region. Errors in κ over this range were larger that those for $\alpha \leq 0.5$.

Optimum pulse widths shifted to longer times as κ was decreased (Figure 11.10). For $\kappa = 5$, small errors in κ were found in the 0.5–20 ms pulsewidth range. Data for 30–50 ms pulses gave very slow convergence of a Marquardt–Levenberg regression program, and the computations usually did not converge even after 120 iterative cycles. Data for $\kappa = 1$ and 0.1 at pulse widths <5 ms showed significant errors in κ. Errors in κ for $\kappa = 0.1$ remained relatively large (15%) even at pulsewidths of 200 ms. At this small value of κ, voltammograms are apparently approaching the *irreversible limit*, beyond which they contain little information about electrode kinetics.

These error analyses clearly show that an optimum range of NPV pulsewidths is required to estimate $k^{o\prime}$ values reliably. If rough estimates of

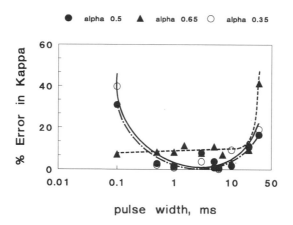

Figure 11.9   Influence of pulsewidth on errors in kinetic parameter for $\kappa = 12$ and three different values of α. Random absolute noise of $\pm 0.5\% i_d$ with $i_d = 1.0$ μA, $E^0 = -0.3$ V, $b_5 = 0.8$ μA V^{-1}, and $b_6 = 0.1$ μA. (Reproduced with permission from [10], copyright by the American Chemical Society.)

$k^{o\prime}$ are obtained before the regression analysis, the appropriate pulsewidth ranges can be chosen from Figures 11.9 and 11.10.

The concept of an optimum time range in electrochemical experiments is a general one. If the time scale is too short or too long for the data to contain sufficient information about the desired kinetic constant, errors will be large. The situation in homogeneous chemical kinetics is similar.

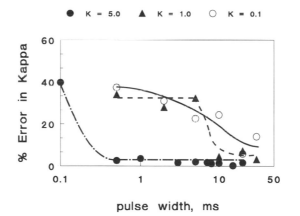

Figure 11.10   Influence of pulsewidth on errors in kinetic parameter for $\kappa = 5$ ($\alpha = 0.35$), 1 ($\alpha = 0.35$), and 0.1 ($\alpha = 0.5$). Random noise of $\pm 0.5\% i_d$ with $i_d = 1.0$ μA, $E^0 = -0.3$ V, $b_5 = 0.8$ μA V^{-1}, and $b_6 = 0.1$ μA. (Reproduced with permission from [10], copyright by the American Chemical Society.)

B.3.d. NPV Analysis of Electron Transfer Rates in Electroactive Films

Kinetics of electron transfer involving electroactive films are important for their characterization and performance [15]. For films of micrometer thickness, diffusion within the film and electron transfer kinetics may influence NPV data. In such cases, it is possible to estimate $k^{o'}$ from NPV data obtained under diffusion–kinetic control. The key requirement is that the diffusion layer developed at the electrode during electrolysis should be a small fraction of the film thickness so that semi-infinite linear diffusion applies [1].

The model in Table 11.8 was used to obtain electrochemical parameters for 20 μm thick films of the protein myoglobin (Mb) and surfactant didodecyldimethylammonium bromide (DDAB) on graphite electrodes. NPV pulse widths of 4–60 ms were used, and the voltammograms featured relatively low signal to noise ratios. From the relation between root mean square displacement Δ^2 and diffusion coefficient [1]

$$\Delta^2 = 2\,D\,t$$

we find for $D = 4 \times 10^{-7}$ cm^2 s^{-1} for Mb-DDAB films [10] the diffusion layer is only 10% of film thickness at the end of a 60 ms pulse and proportionately less at 4 ms. Therefore, the conditions of linear diffusion pertain.

Typical results for Mb-DDAB at small (Figure 11.11) and large (Figure 11.12) pulsewidths show good agreement with the model. Only the rising portion of the curve, between about 2% and 98% of i_d, needs to be analyzed. The small pulsewidth voltammogram is broader than that for the larger pulsewidth because of the increased influence of electrode kinetics at shorter times. Residual plots were nearly random. Relative standard devia-

Table 11.8 Comparison of Electrochemical Parameters of Mb-DDAB Films from NPV and CV[a]

pH/method	$10^7 D'$, cm^{2}s^{-1}	$10^3 k^{o'}$, cm s^{-1}	$-E^{o'}$, V/SCE	α	%RSD[c]
5.5/NPV		7.1 ± 1.5		0.31 ± 0.04	0.26
5.5/NPV	5.7 ± 1.6		0.18 ± 0.05[b]		0.46
5.5/CV	5.1	6.7 ± 0.7	0.124		
7.5/NPV		7.8 ± 1.5		0.27 ± 0.02	0.34
7.5/NPV	4.5 ± 1.1		0.18 ± 0.06[b]		0.53
7.5/CV	3.7	9.0 ± 0.3	0.194		

[a] NPV results obtained on two electrodes by analysis with the model in Table 11.7, see [10]. Values of $k^{o'}$ obtained as average from data at 4–10 ms pulsewidth; D, from data with 30–60 ms pulsewidth.

[b] Estimated as average from 11 data sets with 4–60 ms pulsewidths

[c] Average standard deviation of regression relative to i_d.

the regression-simulation method were not successful because of the difficulty of modeling the large, nonlinear background currents [10].

B.4. Steady State Linear Sweep Voltammograms, Electrochemical Catalysis

Linear sweep voltammetry imposes a linear ramp of potential onto the electrochemical cell in a quiet solution and measures the resulting current vs. potential response (Table 11.1). As in many electrochemical experiments, the shape of the response curve is characteristic of the electrode reaction mechanism [1, 2]. In general, the shapes are complex and not easily described by closed form models.

However, linear sweep voltammetry gives a characteristic shape for electrochemical catalysis if certain experimental conditions are fulfilled. Box 11.2 illustrates a simple pathway for reductive electrochemical catalysis. A catalyst P added to the solution acts as a mediator to shuttle electrons from an electrode to a substrate A, yielding the product B and regenerating P. Substrate of reactant A is directly reducible only at potentials considerably negative of $E^{o'}$. A similar pathway pertains to catalytic oxidations.

The homogeneous second step with rate constant k_1 in Box 11.2 regenerates P in a thin layer of solution near the electrode, so that it can be easily reduced at the electrode. This catalytic cycling of P increases the current and changes the shape of the response curve compared to what would be obtained in the absence of A.

B.4.a. Conventional Size Electrodes

A catalytic reaction can often be made pseudo-first order in Q by using a large excess of the reactant A in the solution. Under this condition, if scan rates (ν) can be achieved such that the reaction in under full kinetic control of the homogeneous chemical reaction with rate k_1, the shape of the LSV response curve is sigmoidal and can be described by the model in Table 11.2. The limiting current (i_l) is related to k_1 by

$$i_l = nFAC_p(D_pC_Ak_1)^{1/2} \tag{11.4}$$

where n is the number of electrons transferred in the catalytic reduction, F is Faraday's constant, A here denotes electrode area, D_P is the diffusion

$$P + e \rightleftharpoons Q \text{ (at electrode), } E^{o'}$$
$$k_1$$
$$Q + A \rightarrow P + B$$

Box 11.2 Simple reductive catalytic pathway.

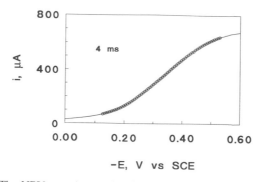

-E, V vs SCE

Figure 11.11 The NPV at a 4 ms pulsewidth of Mb-DDAB film on pyrolytic graphite electrode in pH 5.5 buffer. The line shows experimental NPV; the circles are points computed from a nonlinear regression using the model in Table 11.7 for the parameters found: $k^{o\prime} = 8.1 \times 10^{-3}$ cm s^{-1}, $i_d = 363.7$ μA, $E^{o\prime} = -0.261$ V vs. SCE, $\alpha = 0.28$. (Reproduced with permission from [10], copyright by the American Chemical Society.)

tions of the regression were less than the estimated absolute error in the data of $\pm 0.5\%$ of i_d.

The NPV results for Mb-DDAB films are in good agreement with those obtained from cyclic voltammetry (CV) by conventional peak separation analysis [1, 2]. Nonlinear regression of NPV data for Mb-DDAB films has the advantage over the CV peak separation method of providing a direct test of each voltammogram against the model. Pulsewidths ≤ 10 ms were required for reliable $k^{o\prime}$ values, and longer pulsewidths gave reliable D' values. Nonlinear regression analyses of CV data for Mb-DDAB by using

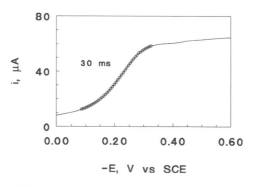

-E, V vs SCE

Figure 11.12 The NPV at a 30 ms pulsewidth of Mb-DDAB film on PG electrode in pH 5.5 buffer. The line shows experimental NPV; the circles are points computed from a nonlinear regression using the model in Table 11.7 for the parameters found: $i_d = 41.3$ μA, $E^{o\prime} = -0.162$ V vs. SCE (κ was not obtainable for this reversible data, see the text). (Reproduced with permission from [10], copyright by the American Chemical Society.)

coefficient of the catalyst, and C_P and C_A are the catalyst and reactant concentrations, respectively. If the limiting current is obtained by fitting the data to the model in Table 11.2, then k_1 can be estimated from eq. (11.4).

The conditions of full kinetic control occur within a given range [1] of k_1/v. If these conditions cannot be attained experimentally, a model based on numerical solutions of the relevant differential equations must be employed.

A requirement for pseudo-first-order conditions is that $C_A \geq 10\ C_P$. Because of the large excess of reactant A, the foot of its direct irreversible reduction wave may overlap the catalytic current–potential curve. In such cases, including an exponential background with the model in Table 11.2 has been successful for the estimation of i_l, which was then used to obtain k_1. Examples of the use of this approach include determination of the rate constant for the reduction of 4,4'-dichlorobiphenyl by anion radicals of anthracenes (Figure 11.13) [16] and by photoexcited organic anion radicals [17].

B.4.b. Ultramicroelectrodes

The model in Table 11.2 using the exponential background term has also been applied to electrochemical catalysis using steady state microelectrode voltammetry. Here, the shape of the wave remains the same at low scan rates irrespective of full kinetic control or pseudo-first-order conditions. The value of i_d estimated by the regression analysis was used to obtain k_1

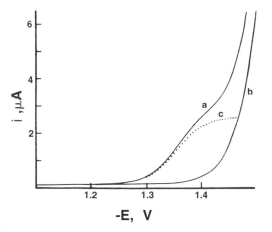

Figure 11.13 Linear sweep voltammogram at a Hg-drop electrode at 0.2 V s^{-1} for the reduction of 4,4'-dichlorobiphenyl (DCB) in 0.1 M tetrabutylammonium iodide in DMF: (a) with 0.1 mM anthracene as catalyst and 4 mM DCB; (b) direct reduction of 4 mM DCB; (c) catalytic component of curve in (a) computed from regression parameters obtained using the model in Table 11.2 with exponential background. (Reproduced with permission from [16], copyright by the American Chemical Society.)

via computation of theoretical curves relating the catalytic limiting current to log k_1 by digital simulation. The reader is referred to the original literature for details [18].

C. Cyclic Voltammetry

C.1. Reversible and Diffusion–Kinetic Control

Cyclic voltammetry is the combination of two linear sweep experiments. The potential is initially swept in a forward direction, and the second scan is the reverse of the first. CV provides signals from the electrolysis of reactants on the forward scan and gives signals from the electrolysis products on the reverse scan [1, 2, 19]. Figure 11.14 shows a typical cyclic voltammogram for a Co(II) complex controlled by electrode kinetics and diffusion. On the forward scan, a peak for the Co(II)/Co(I) reduction is observed. On the reverse scan, the Co(I) that has been formed at the electrode in the forward scan is oxidized back to Co(II).

Cyclic voltammograms (CVs) are rich in information about the mechanism and kinetics of the electrode reaction. Parameters such as D, α, $k^{o'}$, and $E^{o'}$ can be obtained by nonlinear regression analysis of CVs controlled by diffusion and electrode kinetics [5]. CVs can also be employed to examine heterogeneous electron transfer kinetics as well as estimate rate constants (k_{chem}) for a wide variety of chemical reactions coupled to electron transfer steps, such as for the catalytic mechanism in Box 11.2 [5]. Chemical reactions may precede or follow electron transfer at the electrode. Discussion of this vast subject is beyond the scope of this chapter, and the reader is directed to several excellent books [1, 2, 19].

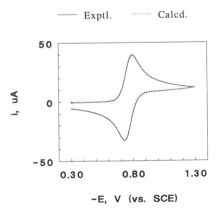

Figure 11.14 Cyclic voltammogram (background subtracted) of 1 mM Co(II)tetraphenyl-porphyrin in 0.1 M TBABr at 25°C showing the results of fitting the data using the CVSIM/CVFIT package.

CVs can be more difficult to analyze by nonlinear regression than steady state voltammograms. First, closed form equations describing the CV responses exist for only a few special cases. In general, numerical simulation techniques must be used to solve the models, which begin as sets of differential equations describing the relevant diffusion and kinetics of the species involved [1, 2]. However, successful solutions to this problem have been devised such that regression analysis using numerically simulated models can now be done on small computers [13, 19, 20].

A more serious problem is that of accounting for the background in CV. The signal to noise ratio is often less than optimum, and background currents may be curved and dependent on a complex set of conditions. Although steady state voltammograms involve background for a single scan, CV background currents occur on both oxidation and reduction cycles. Background subtraction remains a possibility, but background scans on solid electrodes are often not very reproducible.

We are currently aware of three commercially available general programs employing numerically solved or digitally simulated models for nonlinear regression analyses of voltammetric data (Table 11.9). The programs assume linear diffusion and can be used for any system for which this condition applies.

Major advantages of all of the commercial programs are that they allow the analysis of a wide variety of reaction types. The simulation models can generally be used independent of regression analysis. At the time of this writing, limitations in versatility of handling of the backgrounds of CVs existed in all of them. Backgrounds of CVs with irregularly shaped residual currents must be accurately subtracted if good fits are to be obtained. The first program in the list accompanies an excellent book by Gosser [19]. The

Table 11.9 Commercially Available Programs for Digital Simulation–Nonlinear Regression Analyses of Cyclic Voltammograms

Program	Vendor	Capabilities	Limitations[a]
CVSIM/CVFIT	VCH Publishers (with book on CV [19])	Many second-order chemical reactions	Slow for large k_{chem}; slow regression
COOL	PARC[b]	Reliable regression; fast	Pseudo-first-order chemical reactions only
DIGISIM	BAS[c]	Many second-order chemical reactions; fast	Regression convergence poor for very complex mechanisms

[a] Limitations at the time of this writing, which may be corrected in future.
[b] Princeton Applied Research Corp., Princeton, NJ. COOL algorithm also accommodates NPV, SWV, and other digital electrochemical methods.
[c] Bioanalytical Systems, Lafayette, IN.

other two programs are marketed by electrochemical instrument manufacturers.

A comparison of results from commercial programs is illustrated for the CV of Co(II)tetraphenylporphyrin [Co(II)TPP] in N,N-dimethylformamide containing tetrabutylammonium bromide (TBABr) on a glassy carbon electrode. This is a simple one-electron transfer reaction, represented by

$$Co(II)TPP + e \rightleftharpoons Co(I)TPP \qquad k^{o\prime}, E^{o\prime}. \qquad (11.5)$$

Data were successfully fit using a quasireversible model. Results are expressed graphically in Figures 11.14 and 11.15. Although the fits are reasonably good, small systematic deviations of the model and the data can be seen. These deviations are especially apparent on the oxidation scan and may be caused by an inaccurate consideration of the background. As mentioned previously, none of the programs have versatile background models and rely mainly on background subtraction. The situation is not too bad with the Co(II)TPP data, for which S/N is large, but will become more serious for signals that are relatively small with respect to background currents.

Results from different regression-simulation packages agree well. This is illustrated by the parameter values in Table 11.10 for Co(II)TPP, which exhibits reasonably fast electron transfer in DMF. The COOL package gave comparable results.

C.2. Electrochemical Catalysis

As implied in Section B.4.a, the special conditions necessary for steady state catalytic voltammograms in linear sweep or cyclic voltammetry are

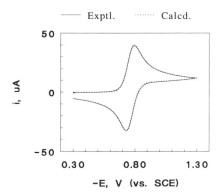

Figure 11.15 Cyclic voltammogram (background subtracted) of 1 mM Co(II)tetraphenylporphyrin in 0.1 M TBABr at 25°C showing the results of fitting the data using the DIGISIM package.

Table 11.10 Parameters Obtained by Nonlinear Regression Analysis of Cyclic Voltammogram of Co(II)TPP at 1 and 5 V s^{-1}

Program	$-E^{o\prime}$, V vs. SCE	α	$k^{o\prime}$ cm s^{-1}	$10^6 D_0$(avg) cm^2 s^{-2}
CVFIT[a]	0.763	0.32	0.088	4.6
DIGISIM[a]	0.763	0.28	0.090	4.6
CVFIT[b]	0.763	0.43	0.059	5.0
DIGISIM[b]	0.763	0.44	0.060	5.0

[a] 1.0 V s^{-1}.
[b] 5.0 V s^{-1}.

not always experimentally attainable. Also, reaction mechanisms for electrochemical catalysis are often more complex than the simple catalytic pathway in Box 11.2.

A method combining nonlinear regression analysis with models computed by digital simulation has been developed for estimating rate constants and analyzing mechanistic nuances for two-electron electrochemical catalysis [21]. The method analyzed forward scans of voltammograms only and includes a linear background term. We illustrate the use of this method by discussing its results for the two-electron catalytic reduction of aryl halides (ArX). The detailed reaction pathway is shown in Box 11.3.

The catalyst P accepts an electron from the electrode in Box 11.3 to begin the reaction. The rate-determining step of the reaction may be either eq. (11.7) or eq. (11.8). Subsequently, several fast reactions (eqs. (11.9) and (11.10), which do not influence the shape of the current–potential curve, give the final hydrocarbon product.

$$P + e \rightleftharpoons Q \text{ (at electrode), } E^{o\prime} \tag{11.6}$$

$$ArX + Q \underset{k_2}{\overset{k_1}{\rightleftharpoons}} ArX\dot{-} + P \tag{11.7}$$

$$ArX\dot{-} \overset{k}{\rightarrow} Ar^{\cdot} + X^{-} \tag{11.8}$$

$$Ar^{\cdot} + Q \rightarrow P + Ar^{-} \tag{11.9}$$

$$Ar^{-} + (H^{+}) \rightarrow ArH \tag{11.10}$$

Box 11.3 Detailed pathway for electrochemical catalytic reduction of aryl halides

A general model based on digital simulation of voltammograms following Box 11.3 was developed to compute the current–potential curves [21]. For analysis of data, the model was considered to represent three separate cases (i) eq. (11.7) as the rate-determining step (rds) (KE or kinetic electron transfer case), (ii) eq. (11.8) as rds (KC or kinetic chemical case), and (iii) mixed kinetic control of both reactions (11.7) and (11.8) (KG or kinetic general case).

Data for the catalytic reduction of 4-chloro- and 4-bromobiphenyl (4-CB and 4-BB) using phenanthridine as a catalyst were analyzed using a regression program employing the Marquardt–Levenberg algorithm. The data were evaluated with the models in the following order: (i) KC model, (ii) KE model, and (iii) KG model. This was done because, when data representing one of the limiting KE or KC cases were fit onto the KG model, serious parameter correlation between rate constants resulted. This occurs because in each limiting case the data contain little information about the rate constant of the step that is not rate determining.

Deviation plots and other criteria of goodness of fit showed that the KE model fit the data best for 4-CB and 4-BB under the experimental conditions used [21]. An example is given in Figure 11.16. Values of k_1 were

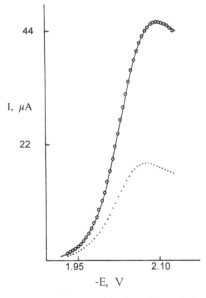

Figure 11.16   Catalytic voltammogram at a Hg-drop electrode for 0.87 mM phenanthridine and 2 mM 4-BB in 0.1 M tetrabutylammonium bromide in DMF. The solid line shows the best fit by regression–simulation onto the KE model and circles are experimental data. The dotted line is voltammogram computed for phenanthridine alone. (Reproduced with permission from [21], copyright by the American Chemical Society.)

$(1.42 \pm 0.12) \times 10^3 M^{-1}s^{-1}$ for the reaction of phenanthridine anion radical (Q) with 4-CB and $(5.1 \pm 1.8) \times 10^4 \ M^{-1}s^{-1}$ for reaction with 4-BB. Both of these values were of similar magnitude but much of better precision than rate constants found under pseudo-first-order conditions.

The inclusion of second-order kinetics in the model allows estimation of much larger rate constants than possible under pseudo-first-order conditions. This is because nearly equal concentrations of reactants can be employed, thus minimizing the rates of very fast reactions compared to pseudo-first-order conditions and making them more accessible to voltammetric measurement. Use of second-order conditions also minimizes contributions to the current from the direct reduction of the reactant. These considerations increase the possible upper limit of k_1 from simulation–regression to $10^8 \ M^{-1}s^{-1}$. The largest rate constant known by the authors to be measured by this method was for the reaction of the Co(I) form of vitamin B_{12} with 1,2-dibromobutene [22], $6.1 \times 10^6 \ M^{-1}s^{-1}$.

D. Square Wave Voltammetry

Square wave voltammetry (SWV) is much more sensitive and has better resolution than cyclic or normal pulse voltammetry. For these reasons, SWV may be the method of choice for systems with low concentrations of electroactive material or with overlapped signals. In this technique (Table 11.1), a square potential pulse of frequency f consisting of a forward and a reverse component is superimposed on top of an increasing potential staircase and applied to the electrochemical cell (Figure 11.17)[13]. Measurements of current are timed to occur at the end of each forward and reverse pulse. These two currents are subtracted to give a difference current

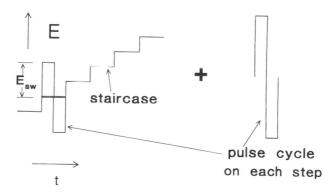

Figure 11.17 Square wave voltammetry (SWV) input waveform separated into its components parts of a staircase with superimposed forward–reverse square pulse cycle superimposed on top of each step.

that has a very large faradaic to charging current ratio. The difference current output is displayed vs. the potential of the step on the staircase. The difference current response curve for a reversible electrochemical reaction is a symmetric peak, contrasting to the waves and unsymmetric peaks obtained in other voltammetric methods.

As with CV, closed form models for SWV are rarely available. However, the COOL program mentioned in Table 11.9 was designed by Janet and R. A. Osteryoung and their coworkers to analyze SWV and other types of pulse voltammetry data [13]. We illustrate this by an application of the COOL algorithm to SWV data for the reduction of Zn^{2+} to $Zn(Hg)$ in M KNO_3 at a mercury electrode using a diffusion–kinetic model [12]. Excellent fits were obtained (Figure 11.18). Parameters and their confidence intervals were $k_a^o = (k^{o\prime}) = 2.64 \pm 0.16 \times 10^{-4}$ cm/s, $\alpha = 0.20 \pm 0.02$, and $E_{1/2}^r = 1.000 \pm 0.001$ V vs. SCE, illustrating the precision available from the method.

Figure 11.18 shows the shapes of SWV difference current curves under experimental conditions where the reduction of Zn^{2+} is reversible (10 Hz) and nearly irreversible (500 Hz). The curve attains an overlapped peak appearance at the higher frequencies as a consequence of the subtraction of the forward and reverse currents [12, 13].

As in CV, a variety of coupled chemical reactions under pseudo-first-order conditions can be investigated by using the COOL program to analyze SWV data. A considerable number of electrode reactions of different types have been analyzed using this approach [13].

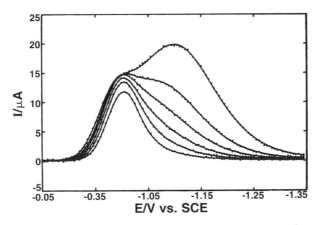

Figure 11.18 Square wave voltammograms for reduction of 1 mM Zn^{2+} in M KNO_3 at $E_{step} = 5$ mV and $E_{sw} = 25$ mV. The points show experimental data, and the solid lines show the best fits to a diffusion–kinetic model using the COOL algorithm. The frequency in ascending order of curves is 10, 25, 100, 200, 500 Hz. (Adapted with permission from [12], copyright by Elsevier.)

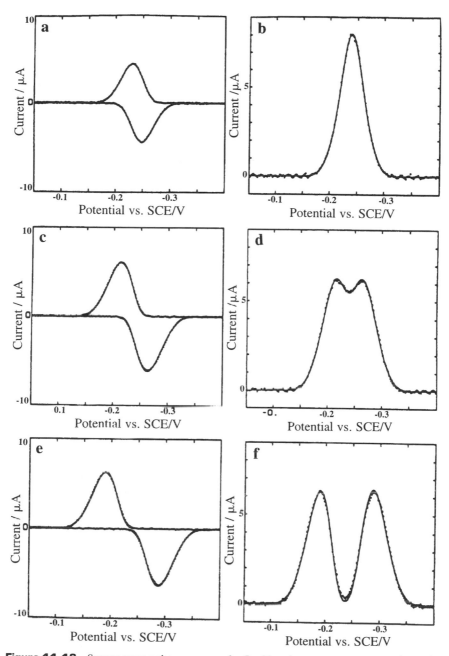

Figure 11.19 Square wave voltammograms for 5 mM azobenzene on a mercury electrode at $f = 200$ Hz. Left side depicts forward (positive) and reverse (negative) current, and the corresponding difference current curves are shown on the right side. $E_{step} = 5$ mV and $E_{sw} = 25$ (A, B), 50 (C, D) and 75 (E, F) mV. The points show experimental data, and the solid lines show the best fits to an adsorption–kinetic model using the COOL algorithm. (Adapted with permission from [23], copyright by the American Chemical Society.)

The COOL algorithm has also been applied to adsorbed molecules on electrodes and can be used to obtain kinetic and thermodynamic parameters for such systems [23]. The model predicts the variety of difference current peak shapes that can be encountered for an adsorbed system. The shapes depend on $k°t_p$, where $k°$ is the first-order electron transfer rate constant in s^{-1} and t_p is the pulse width, which is inversely proportional to frequency. The shapes also depend on the square wave pulse height, E_{sw}. These different shapes are illustrated by SWV data for azobenzene adsorbed on mercury (Figure 11.19). The forward and reverse current curves on the left give rise to the difference current curves on the right-hand side of the figure. At a pulse height E_{sw} of 25 mV, a single peak is found, but at E_{sw} of 75 mV two peaks result. This is a consequence of the subtraction of the forward and reverse currents, as in the case of reduction of Zn^{2+}.

The goodness of fit of the adsorption model to the data for azobenzene on mercury is also illustrated in Figure 11.19. The azobenzene surface concentration, electrochemical transfer coefficient, and electron transfer rate constant were obtained by using the COOL algorithm. In 5 μM solutions of azobenzene in acetate buffer, the surface concentration on mercury was 26 pmol cm^{-2}, the standard potential was -0.239 V vs. SCE, the transfer coefficient was 0.51 and the rate constant was 160 s^{-1}.

Because the charging current component is nearly eliminated from the difference current, SWV alleviates some but not all of the problems with background encountered in CV. However, as for CV, there is presently no way to account for a sloping or otherwise structured background within the COOL program package if background subtraction is not effective.

References

1. A. J. Bard and L. R. Faulkner, *Electrochemical Methods.* New York: Wiley, 1980.
2. P. H. Reiger, *Electrochemistry.* Englewood Cliffs, N.J.: Prentice-Hall, 1987.
3. Meites, L., "Some New Techniques for the Analysis and Interpretation of Chemical Data," *CRC Critical Reviews in Anal. Chem.* 8 (1979), pp. 1–53.
4. M. Fleischmann, S. Pons, D. R. Rolison, and P. P. Schmidt, *Ultramicroelectrodes.* Morganton, N.C.: Datatech, 1987.
5. J. F. Rusling, "Analysis of Chemical Data by Computer Modeling," CRC Critical Reviews in Anal. Chem. 21 (1989), pp. 49–81.
6. A. Owlia and J. F. Rusling, "Nonlinear Regression Analysis of Steady-State Voltammograms Obtained at Microelectrodes," *Electroanalysis* 1 (1989), p. 141.
7. A. Owlia, Z. Wang, and J. F. Rusling, "Electrochemistry and Electrocatalysis with Vitamin B_{12} in an AOT Water-in-Oil Microemulsion." *J. Am. Chem. Soc.* 111 (1989), pp. 5091–5098.
8. R. N. Adams, *Electrochemistry at Solid Electrodes.* New York: Marcel Dekker, 1969.
9. G. E. Cabannis, A. A. Diamantis, W. R. Murphy, R. W. Linton, and T. J. Meyer, "Electrocatalysis of Proton-coupled Electron-Transfer Reactions at Glassy Carbon Electrodes," *J. Am. Chem. Soc.* 107 (1985), p. 1845.

10. J. F. Rusling and A.-E. F. Nassar, "Electron Transfer Rates in Electroactive Films from Normal Pulse Voltammetry. Myoglobin-Surfactant Films," *Langmuir* **10** (1994), 2800–2806.

11. G. N. Kamau, W. S. Willis, and J. F. Rusling, "Electrochemical and Electron Spectroscopic Studies of Highly-Polished Glassy Carbon Electrodes," *Anal. Chem.* **57** (1985), p. 545.

12. W. S. Go, J. J. O'Dea, and J. Osteryoung, "Square Wave Voltammetry for the Determination of Kinetic Parameters," *J. Electroanal. Chem.* **255** (1988), p. 21.

13. J. G. Osteryoung and J. J. O'Dea, "Square Wave Voltammetry," in A. J. Bard (Ed.), *Electroanalytical Chemistry*, vol. 14, pp. 209–308. New York: Marcel Dekker, 1986.

14. K. B. Oldham, "Approximations for the $x \exp x^2 \operatorname{erfc} x$ Function," *Math Comput.* **22** (1968), p. 454.

15. Murray, R. W. (Ed.), *Molecular Design of Electrode Surfaces, Techniques in Chemistry*, vol. 22. New York: Wiley, 1992.

16. J. F. Rusling and T. F. Connors, "Determination of Rate Constants of Pseudo-First-Order Electrocatalytic Reactions from Overlapped Voltammetric Data," *Anal. Chem.* **55** (1983), pp. 776–781.

17. S. S. Shukla and J. F. Rusling, "Photoelectrocatalytic Reduction of 4-Chlorobiphenyl Using Anion Radicals and Visible Light," *J. Phys. Chem.* **89** (1985), pp. 3353–3358.

18. C. L. Miaw, J. F. Rusling, and A. Owlia, "Simulation of Two-Electron Homogeneous Electrocatalysis for Steady State Voltammetry at Hemispherical Microelectrodes," *Anal. Chem.* **62** (1990), pp. 268–273.

19. D. Gosser, *Cyclic Voltammetry*, New York: VCH Publishers, 1994.

20. M. Rudolph, D. P. Reddy, and S. W. Feldberg, "A Simulator for Cyclic Voltammetric Responses," *Anal. Chem.* **66** (1994), pp. 589A–600A.

21. J. V. Arena and J. F. Rusling, "Kinetic Control in Two-Electron Homogeneous Redox Electrocatalysis. Reduction of Monohalobiphenyls," *J. Phys. Chem.* **91** (1987), pp. 3368–3373.

22. T. F. Connors, J. V. Arena, and J. F. Rusling, "Electrocatalytic Reduction of Vicinal Dibromides by Vitamin B_{12}," *J. Phys. Chem.* **92** (1988), pp. 2810–2816.

23. J. J. O'Dea and J. Osteryoung, "Characterization of Quasi-Reversible Surface Processes by Square-Wave Voltammetry," *Anal. Chem.* **65** (1993), p. 3090.

Chapter 12

Chronocoulometry

A. Basic Principles

Chronocoulometry is a transient electrochemical technique in which a single or double pulse of potential is applied to an electrolytic cell (Figure 12.1). The resulting response curve is the cumulative charge (Q) passing through the cell measured vs. time. As in voltammetry, cells contain reference and counter electrodes and a working electrode. The reference electrode is designed to hold a constant potential, and a potentiostatic circuit is used so that a variation in the potential applied to the cell changes the oxidizing or reducing power of the working electrode. With this basic arrangement, electrochemical events occurring at the working electrode are reflected in the charge vs. time data obtained.

For the simple electrode reaction

$$O + e^- \rightleftharpoons R \qquad E^{o'} \tag{12.1}$$

the typical form of single and double potential steps (Figure 12.1) shows that an intial potential (E_i) is chosen to be positive of $E^{o'}$, and the final potential of the first pulse is far enough negative of $E^{o'}$ that rapid, diffusion-controlled electrolysis takes place [1]. In a double potential step experiment, the second pulse often returns to the initial potential.

The charge passing through an electrochemical cell is the integral of the current vs. time response. That is,

$$Q = \int_0^t i \, dt. \tag{12.2}$$

This integral may be obtained electronically or computed by numerical integration of the current–time data. The input potential waveform in chronocoulometry is identical to the that in chronoamperometry. The only difference between the techniques is that chronocoulometry outputs the charge, whereas chronoamperometry outputs current. Models for chrono-

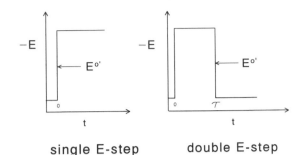

single E-step   double E-step

Figure 12.1   Potential pulse waveforms used in chronocoulometry.

coulometry can be obtained from integration of the models for chrono-amperometry [1].

Chronocoulometry has certain distinct advantages over chronoampero-metry. The large initial charging current at the electrode upon stepping the potential in chronoamperometry can often be treated as a constant charge in chronoamperometry. Since the current is integrated to obtain the charge, the signal to noise ratio is better in chronocoulometry because of averaging of the random noise in the current. For these reasons, chronocoulometry is generally the preferred approach for parameter estimation by nonlinear regression. We deal with only this variant of the potential step experiment.

We assume that chronocoulometry will be used for the estimation of accurate and precise values of parameters of an electrochemical system that has been investigated at least qualitatively by other methods. It would be unusual to employ chronocoulometry to a completely unknown electrode reaction without having used other techniques to investigate the system first. Candidate systems would typically be examined first by cyclic or square wave voltammetry. Chronocoulometry might then be used to obtain the diffusion coefficient of an electrochemical reactant, the surface concentration of an adsorbed electroactive species, or a rate constant for a step in the electrode reaction pathway. The following sections deal with these topics in turn.

B. Estimation of Diffusion Coefficients

B.1. Reversible Electron Transfer—Single Potential Step

In principle, it should be possible to pulse the potential to values for which conversions of O to R, or R to O, are rapid and diffusion controlled. The diffusion coefficient can be determined from a single potential step

Table 2.2 Linear Regression Analysis of Beer's Law Data

x_j (μM)	y_j(meas) (A)	y_j(calc) (A)	Difference y_j(meas) $-$ y_j(calc)	Diff./SDR[a]
1.0	0.011	0.01118	−1.84E-04	−0.0754
5.0	0.049	0.05096	−1.96E-03	−0.803
10.0	0.102	0.10068	1.32E-03	0.541
20.0	0.199	0.20013	−1.13E-03	−0.463
30.0	0.304	0.29957	4.43E-03	1.81
40.0	0.398	0.39901	−1.01E-03	−0.414
50.0	0.497	0.49846	−1.45E-03	−0.595

Slope = 9.944367E-03
Standard deviation of the slope = 5.379905E-05
Intercept = 1.239661E-03
Standard deviation of the intercept = 1.51158E-03
Correlation coefficient (**r**) = 0.9999269
[a] Standard deviation of the regression = 2.438353E-03

standard deviation of the regression (SDR) is expressed in the units of y. To interpret this statistic in terms of goodness of fit, we compare it to the standard error (e_y) in measuring y. In this case, e_y is the standard error in A. Suppose we use a spectrophotometer that has an error in the absorbance $e_A = \pm0.003$, which is independent of the value of A. (This error would not be very good for a modern spectrophotometer!) We have SDR = 0.0024 from Table 2.2. Therefore, SDR < e_A, and we consider this as support for a good fit of the model to the data.

Although summary statistics such as the correlation coefficient and SDR are useful indicators of how well a model fits a particular set of data, they suffer from the limitation of being *summaries*. We need, in addition, to have methods that test each data point for adherence to the model. We recommend as a general practice the construction of graphs for further evaluation of goodness of fit. The first graph to examine is a plot of the experimental data along with the calculated line (Figure 2.1(a)). We see that good agreement of the data in Table 2.2 with the straight line model is confirmed by this graph. A very close inspection of this type of plot is often needed to detect systematic deviations of data points from the model.

A second highly recommended graph is called a *deviation plot* or *residual plot*. This graph provides a sensitive test of the agreement of individual points with the model. The residual plot has $[y_j(\text{meas}) - y_j(\text{calc})]$/SDR on the vertical axis plotted against the independent variable. Alternatively, if data are equally spaced on the x axis, the data point number can be plotted on the horizontal axis [1]. The quantities $[y_j(\text{meas}) - y_j(\text{calc})]$/SDR SDR are called the *residuals* or *deviations*. They are sometimes given the symbols dev_j. The dev_j are simply the differences of each experimental

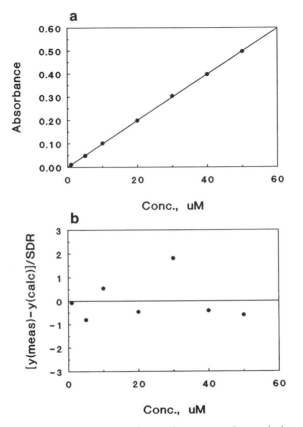

Figure 2.1 Graphical output corresponding to linear regression analysis results in Table 2.2: (a) points are experimental data, line computed from regression analysis; (b) random deviation plot.

data point from the calculated regression line normalized by dividing by the standard deviation of the regression. Division by SDR allows us to plot all deviation plots on similar y-axis scales, which are now scaled to the standard deviation in y. This facilitates comparisons of different sets of experimental data.

In the example from Table 2.2, the residual plot (Figure 2.1(b)) has points randomly scattered around the horizontal line representing $dev_j = 0$. This random scatter indicates that the model provides a good fit to the data within the confines of the signal to noise ratio of the measurements. This type of plot is a highly sensitive indicator of goodness of fit.

So far, we have discussed only data that agree very well with the chosen model. We now explore a second set of data that are fit less well by the linear Beer's law model. The data in Table 2.3 gave a slope of

Table 12.1 Model for Reversible Electron Transfer in Single Potenti
Chronocoulometry [2, 3]

Assumptions: Reversible electron transfer, linear and edge diffusion
Regression equation:

$$Q(t) = b_0 + b_1 t^{1/2} + b_2 t$$
$$b_o = Q_{dl}; \qquad b_1 = 2FAD^{1/2}C\pi^{-1/2};$$
$$b_2 = aFADC(\pi r)^{-1/2}$$

Q_{dl} = double layer charge; F = Faraday's constant; A = electrode area;
   coefficient; C = concentration of reactant; a = constant accounting for nonlin
   and r = radius of disk or spherical electrode

Regression parameters: *Data:*
   $b_0 \quad b_1 \quad b_2$ $Q(t)$ vs. t

Special instructions: Compute D from b_1: background, especially on solid electrc
   need subtraction

---

experiment of this sort. The semi-infinite linear diffusion model for a :
disk or mercury drop (spherical) electrode with a correction for nonlir
diffusion is given in Table 12.1. Although linearization of the model is of
used to analyze chronocoulometric data, this means that the nonline
diffusion term must be ignored, which can lead to significant errors [2, 4

The characteristic response of the model in Table 12.1 is shown in Figur
12.2. The single-step response is shown along with that for the reversible
double potential step experiment to be discussed later. In this chapter, we
shall consider the charge generated on the forward potential pulse to be

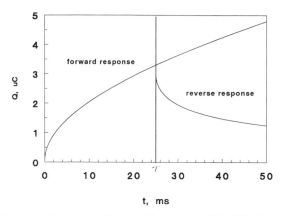

Figure 12.2 Theoretical responses for an arbitrary reversible, diffusion-controlled electrochemical reaction. Forward response for single potential step is shown over 50 ms. Forward
and reverse responses are shown for a double potential step and a switching time (τ) of
25 ms. In the latter case, the forward response is observed for only 25 ms, where the reverse
response begins.

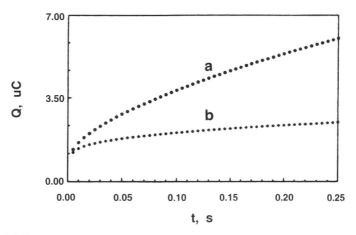

Figure 12.3 Single potential step chronocoulometry of 1 mM vitamin B$_{12r}$ in pH 2.4 phosphate buffer in water/acetonitrile (1:1) E_i = −0.35 V vs. SCE; E_f = −1.0 V vs. SCE: (a) raw data for 1 mM vitamin B$_{12r}$; (b) background charge for solution not containing vitamin B$_{12r}$. (Reprinted with permission from [3], copyright by VCH.)

Table 12.2, we see that when the raw chronocoulometric data were analyzed by using the model in Table 12.1, a value of the diffusion coefficient (D) much larger than that obtained by cyclic voltammetry was obtained. However, when the background subtracted data for vitamin B$_{12r}$ were analyzed, a value similar to that from CV was obtained. An excellent fit of the reversible model to the data was obtained (Figure 12.4). This example

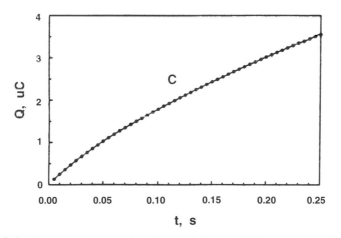

Figure 12.4 Chronocoulometry of 1 mM vitamin B$_{12r}$ in pH 2.5 phosphate buffer in water/ acetonitrile (1:1) after background subtraction. E_i = −0.35 V vs. SCE; E_f = −1.0 V vs. SCE. Points are experimental data; line results from fit onto reversible model, Table 12.1. (Reprinted with permission from [3], copyright by VCH.)

illustrates the importance of background subtraction when using solid electrodes with chronocoulometry to obtain diffusion coefficients.

B.2. Reversible Electron Transfer—Double Potential Step

Double potential step chronocoulometry can also be used to obtain diffusion coefficients. The model is outlined in Table 12.3, and the reversible response is shown in Figure 12.2. An advantage is that the double-step data provides a better test for reversibility than the rather featureless single-step signal. It also provides additional information for model testing in cases where the reversible model might not apply. A disadvantage for systems with large background currents is that there is ambiguity about exactly how to account for the background.

The double potential step method works quite well on pure mercury electrodes, on which the background current is predominantly the double layer charge Q_{dl}. D values for several metal ions showed excellent precision and good agreement with values obtained from dc polarography (Table 12.2) [2].

B.3. Diffusion–Kinetic Model for a Single Potential Step

When electron transfer is not fully at equilibrium during the time scale of the experiment, the shape of the $Q(t)$ vs. t curve is controlled by electrode kinetics and diffusion. Analysis of such data by nonlinear regression analy-

Table 12.3 Model for Reversible Electron Transfer in Double Potential Step Chronocoulometry [2]

Assumptions: Reversible electron transfer, linear and edge diffusion
Regression equation:
$$Q(t < \tau) = b_0 + b_1 t^{1/2} + b_2 t$$
$$Q(t < \tau) = b_1[t^{1/2} - (t - \tau)^{1/2}] + b_2\tau + b_3(\tau - t)$$
$$b_0 = Q_{dl}; \qquad b_1 = 2FAD_O^{1/2}C\pi^{-1/2};$$
$$b_2 = aFAD_OC(\pi r)^{-1/2}; \qquad b_3 = aFAD_RC(\pi r)^{-1/2}$$

Q_{dl} = double layer charge; F = Faraday's constant, A = electrode area; D_O and D_R = diffusion coefficients; C = concentration of reactant; a = constant accounting for nonlinear diffusion; and r = radius of disk or spherical electrode; model adaptable to $D_O \neq D_R$

Regression parameters: *Data:*
$b_0 \quad b_1 \quad b_2 \quad b_3$ $Q(t)$ vs. t

Special instructions: Logical statements required in subroutine for computing $Q(t)$:
IF $t < \tau$ THEN (use) $Q(t) = Q(t < \tau)$
IF $t > \tau$ THEN (use) $Q(t) = Q(t > \tau)$
Background may need subtraction or consideration in the model for solid electrodes; obtain D_O from b_1.

Table 12.4 Model for Diffusion–Kinetic Control in Single Potential Step Chronocoulometry [3]

Assumptions: Kinetic and linear diffusion control of Q, no adsorption

Regression equation:

$$Q(t) = b_1[\exp(y^2)\, \text{erf}(y) + 2\pi^{-1/2} - 1]$$

$$y = b_0^{1/2}; \qquad b_0 = \lambda; \qquad b_1 = \frac{\theta'}{b_0^2}$$

$$\theta' = nFAK^{O'}\exp\{-\alpha(F/RT)n(E-E^{O'})\}$$

$$\lambda = k^{O'}[D_O^{-1/2}\exp\{-\alpha(F/RT)n(E-E^{O'})\} + D_R^{-1/2}\exp\{-\alpha(F/RT)n(E-E^{O'})\}]$$

$F =$ Faraday's constant, $A =$ electrode area; D_R and $D_O =$ diffusion coefficients; $C =$ concentration of reactant; $T =$ temperature in Kelvins.

Regression parameters:			*Data:*
b_0	b_1	b_2	$Q(t)$ vs. t

Special instructions: Estimation of D and $k^{O'}$ require values of $E^{O'}$ and α to be known or determined from a voltammetric method; background requires subtraction; compute $\exp(y^2)\,\text{erf}(y)$ from standard series expansion [5]

---

sis provides the diffusion coefficient simultaneously with the apparent standard heterogeneous rate constant for electron transfer. The model is called the *diffusion–kinetic model* (Table 12.4).

The diffusion–kinetic model has been applied to the oxidation of ferrocyanide ion:

$$\text{Fe(CN)}_6^{4-} \rightleftharpoons \text{Fe(CN)}_6^{3-} + e^-. \tag{12.4}$$

In neutral 0.1 M KCl this redox couple showed an anodic voltammetric peak at about 0.25 V vs. SCE. Starting from initial potentials of -0.2 V, single potential step chronocoulometry was done with final potentials of 0.4, 0.5, and 0.6 V vs. SCE. These final potentials were well positive of the $E^{o'}$ of this redox couple as judged from the midpoint of the anodic and cathodic peaks in the cyclic voltammogram. When the reversible model in Table 12.1 was fit to the background subtracted data, nonrandom deviation plots (Figure 12.5) and a D about 15% larger than the accepted value were obtained [3].

The final potential used in these experiments should have been positive enough to ensure adherence of the data to the conditions of the reversible model. However, the model failed to fit the data. It was suspected that an inorganic polymer film formed on the electrode during the chronocoulometry may have partially blocked the surface and slowed down electron transfer [3].

Analysis of the background-subtracted ferrocyanide Q-t data with the diffusion–kinetic model in Table 12.4 showed all the characteristics of a good fit (Figure 12.6), including a random deviation plot. The average D of $(6.3 \pm 0.15) \times 10^{-6}$ cm^2 s^{-1} was in good agreement with the value from the literature of 6.5×10^{-6} cm^2 s^{-1}. The $k^{o'}$ of $(6 \pm 2) \times 10^{-3}$ cm s^{-1} at pH 7 (7×10^{-3} cm s^{-1} was found by CV) decreased to $(2.0 \pm 0.9) \times 10^{-4}$ cm

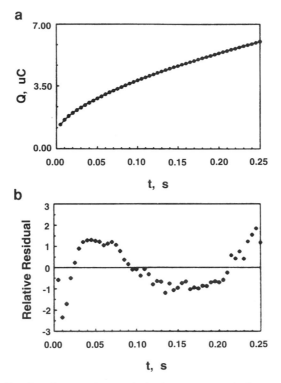

Figure 12.5 Results of regression analysis for uncorrected chronocoulometric data (E_i = −0.20 V vs. SCE; E_f = 0.40 V vs. SCE) for 0.2 mM ferrocyanide in 0.1 M KCl, pH 12: (a) experimental charge (○); line from fit onto reversible model; (b) plot of relative residuals from the analysis. (Reprinted with permission from [3], copyright by VCH.)

s^{-1} at pH 12, in line with the hypothesis of film formation on the electrode, which is facilitated by alkaline solutions [3]. This example illustrates the importance of choosing the correct model for the system to obtain accurate parameters.

C. Surface Concentrations of Adsorbates from Double Potential Steps

This model describes an electrochemical reactant that can be electrolyzed both while adsorbed on the electrode and by diffusion through the solution to the electrode. This can be expressed by the electron transfer reactions in Box 12.1. Both O and R are adsorbed on the electrode, but O is adsorbed to a greater extent than R. The model as used for nonlinear regression is presented in Table 12.5.

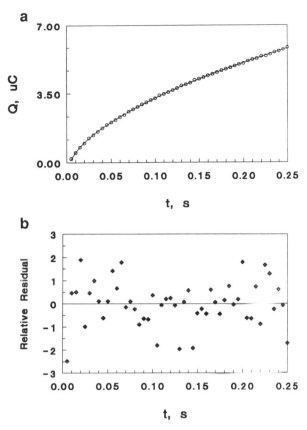

Figure 12.6 Results of regression analysis for background subtracted chronocoulometric data ($E_i = -0.20$ V vs. SCE; $E_f = 0.40$ V vs. SCE) for 0.2 mM ferrocyanide on 0.1 M KCl, pH 12: (a) background subtracted experimental charge (○); line from fit onto diffusion–kinetic model (Table 12.4); (b) plot of relative residuals from the analysis. (Reprinted with permission from [3], copyright by VCH.)

The surface concentration of O (Γ_0) is of prime importance in studies of adsorption on electrodes. Pointwise analysis of variance showed that, for values of t > 50 ms, the most information about Γ_0 is contained in the first 25 ms after $t = 0$, and the first 25 ms after the potential switch at $t = \tau$

$$O + e^- \rightleftharpoons R \qquad E^{o'} \qquad (12.5)$$
$$O_{ads} + e^- \rightleftharpoons R_{ads} \qquad E_{ads}^{o'}. \qquad (12.6)$$

Box 12.1 Reactant adsorption, reversible electron transfer.

Table 12.5 Model for Reversible Electron Transfer with Reactant Adsorption in Double Potential Step Chronocoulometry [2]

Assumptions: Reversible electron transfer, linear + edge diffusion; $D_O \neq D_R$; reactant adsorbed, product dissolves in solution (can modify for reactant + product adsorbed) [2]; electrolysis of adsorb species and double layer charging occur instantaneously

Regression equation:

$$Q(t < \tau) = b_0 + b_1 t^{1/2} + b_2 t$$
$$Q(t < \tau) = b_1[t^{1/2} - (t - \tau^{1/2}] + b_2\tau + b_3(\tau - t) + b_4(2/\pi) \sin^{-1}(t/\tau)^{1/2}$$
$$b_0 = Q_{dl} + nFA\Gamma_0; \qquad b_1 = 2nFAD^{1/2}C\pi^{-1/2};$$
$$b_2 = aFAD_OC(\pi r)^{-1/2}; \qquad b_3 = aFAD_RC(\pi r)^{-1/2}; \qquad b_4 = nFA\Gamma_0$$

Q_{dl} = double layer charge; F = Faraday's constant; A = electrode area; D_O, D_R are diffusion coefficients; C = concentration of reactant; a = constant accounting for nonlinear diffusion; and r = radius of disk or spherical electrode

Regression parameters: *Data:*

b_0 b_1 b_2 b_3 b_4 $Q(t)$ vs. t

Special instructions: Logical statements required in subroutine for computing $Q(t)$:

> IF $t < \tau$ THEN (use) $Q(t) = Q(t < \tau)$
> IF $t > \tau$ THEN (use) $Q(t) = Q(t > \tau)$

Use $\tau \leq 25$ ms for accurate Γ_0 estimates [2]; for solid electrodes, background may need subtraction or consideration in the model; obtain Γ_0 from b_4

---

[2]. When data are equally spaced on the t axis, accurate estimates of Γ_0 require $\tau \leq 25$ ms. Data with larger values of τ can be analyzed successfully by using an analysis schedule that includes data grouped in the regions of greatest significance; that is, immediately after $t = 0$ and $t = \tau$.

The model in Table 12.5 was used to analyze data for Cd(II) ions adsorbed onto mercury electrodes from aqueous sodium thiocyanate/sodium nitrate solutions [2]. Thiocyanate induces adsorption of Cd(II), and both adsorbed and diffusing Cd(II) are reduced to Cd(Hg). When 50-data point sets using only the time regions of highest Γ_0 information content were analyzed, excellent fits were obtained (Figure 12.7). Agreement of D and Γ_0 was found [2] with a previous study done using a linear plot approximation [6]. Some of the results are summarized in Table 12.6. When 50-data point sets were analyzed for $\tau > 100$ ms with equal spacing on the t axis, large systematic errors in Γ_0 resulted from both nonlinear regression and linear plot methods.

The residual plot obtained from fitting an unknown set of chronocoulometric data with the reversible model in Table 12.3 can be used as a test for adsorption. Figure 12.8 shows such a plot resulting from a fit of data for Cd(II) on a mercury electrode onto the reversible model. The experimental points fall along the lines for the theoretical deviation plot obtained by fitting data computed from the adsorption model onto the reversible model. The characteristic shape of this deviation plot suggests the presence of adsorbed reactant. This can be confirmed by fitting the data with the adsorp-

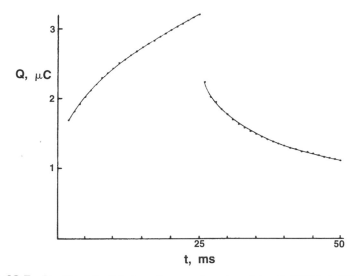

Figure 12.7 Double potential step chronocoulometry of 1 mM Cd(II) in 0.8 NaNO3/ 0.2 M NaNO₃: experimental charge (●); line from fit onto adsorption model (Table 12.5). (Reprinted with permission from [2], copyright by the American Chemical Society.)

tion model in Table 12.5. The resulting residual plot from such an analysis is random (Figure 12.9), confirming the hypothesis suggested by Figure 12.8.

D. Rate Constant for Reaction of a Product of an Electrochemical Reaction

D.1. Chronocoulometry in Studies of Reactive Species

Electron transfer reactions at electrodes can yield reactive intermediates such as radicals and ion radicals. Electrochemical methods such as chronocoulometry can be used to generate reactive species and study their reac-

Table 12.6 Surface Concentrations and Diffusion Coefficients of Cd(II) from Analysis of Double Potential Step Data [3] Using the Model in Table 12.5

Parameter	$\tau = 25$ ms	$\tau = 100$ ms[a]	lit. [ref.]
$10^6 D$, cm²s⁻¹	8.7 ± 0.3	7.4 ± 0.3	7.4 [7], 7.8 [2]
$nF\Gamma_0$, µC cm⁻²	24.2 ± 1.1	23.7 ± 2.1	21.1 [7]

[a] Analysis of one data point/ms in regions $t = 4–30$ ms and $94–116$ ms.

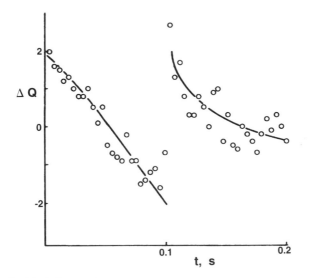

Figure 12.8 Residual plot test for reactant adsorption. The solid line shows the theoretical residual (deviation) pattern for noise-free data following the model in Table 12.5 fit onto the reversible model in Table 12.3. The points represent residual plot from analysis of data in Figure 12.7 with the reversible model. (Reprinted with permission from [2], copyright by the American Chemical Society.)

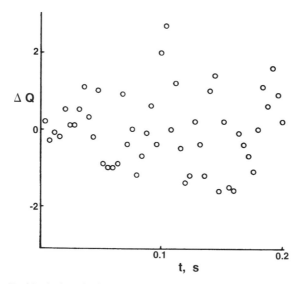

Figure 12.9 Residual plot obtained after regression analysis represented in Figure 12.7 onto the adsorption model in Table 12.5. (Reprinted with premission from [2], copyright by the American Chemical Society.)

tions. Depending on the exact sequence of events in the mechanism of the electrode reaction, the shape of the chronocoulometric response curve will change.

For example, suppose the product of an electrode reaction reacts with water or the solvent (S) to form an electroinactive species. The reaction pathway can be expressed by the equations in Box 12.2.

In Box 12.2, the reactive intermediate B formed at the electrode (eq. (12.7)) undergoes a chemical reaction with S in the solution (eq. (12.8)). Electrochemists call the pathway in Box 12.2 an *EC reaction*, because an electron-transfer step (E) is followed by a chemical step (C). For such a reaction, if k_1 is very large relative to $1/\tau$, the charge on the reverse pulse will not decrease as it would in a reversible electron transfer, because no B is present to oxidize. All the B has reacted (eq. (12.8)). The charge on the forward pulse will appear identical to a reversible reaction. If $k_1\tau$ is in the correct range, a partial decrease of charge relative to a reversible reaction will be observed. These data can be analyzed to estimate k_1. A wide variety of electrochemical reaction types have been found, and models for many of them have been published [1, 7].

D.2. Chemical Rate Constants of ECE-Type Reactions and Distinguishing between Alternative Pathways

Analysis of data reflecting the EC reaction mechanism just discussed requires double potential step chronocoulometry, because only the response during the reverse step is influenced by the chemical reaction. In many cases, however, the product of the chemical step can be electrolyzed. This leads to additional charge on the forward potential step above that expected for an EC reaction. Such an electrode reaction pathway is called an *ECE mechanism* (Box 12.3), in which the EC process is followed by a second electron transfer.

Although the electrode reaction mechanism of an ECE process may be complex, the data on the forward pulse of a potential step reflect the rate of the chemical reaction. This is because, when the potential step (for oxidations) is well positive of both formal potentials for A/B and C/D couples, the electrolysis of C provides additional charge over that obtained from only the first electron transfer reaction. The amount of additional

---

$$A \underset{}{\overset{}{\rightleftharpoons}} B + e^- \qquad\qquad (12.7)$$
$$B + S \xrightarrow{k_1} C \qquad\qquad (12.8)$$

Box 12.2 EC reaction, reversible electron transfer.

$$A \underset{k}{\rightleftharpoons} B + e^- \qquad E^{o'}_1 \qquad (12.9)$$
$$B \xrightarrow{k} C \qquad\qquad (\text{rds}) \qquad (12.10)$$
$$C \rightleftharpoons D + e^- \qquad E^{o'}_2. \qquad (12.11)$$

Box 12.3 ECE pathway, reversible electron transfer.

charge is proportional to the rate of transformation of B to C. Thus, single potential step chronocoulometry can be used to find the rate constant of the chemical reaction. The data analysis problem is a bit simpler than for a double potential step experiment.

We now discuss analysis of data on ECE reactions. Equation (12.10) is written as a pseudo-first-order reaction to simplify the model. In general, we would have

$$B + S \xrightarrow{k_2} 2 C. \qquad (12.12)$$

The usual pseudo-first-order condition for ECE reactions is $[S] > 10\times[A]$. If this can be achieved experimentally, with $k\tau$ in a range where the charge passed represents less then 2F/mol of A, then k can be estimated by nonlinear regression analysis. The second-order rate constant is found from $k_2 = k/[S]$. The pulse width τ must be chosen correctly to achieve the above conditions [8].

As discussed earlier, if $E_{\text{pulse}} \gg E^{o'}_1 \gg E^{o'}_2$, the intermediate C will be oxidized as soon as it is formed. The electron can be lost either at the electrode (eq. (12.11)) or in a thermodynamically favored chemical disproportionation step (eq. (12.13), Box 12.4) [9]. This latter pathway is featured in the so-called DISP1 mechanism shown in box 12.4, which has eq. (12.10) as rate-determining step.

Actually, the general mechanism for such reactions includes both Boxes 12.3 and 12.4, and the possibility of eq. (12.13) as a rate-determining step (rds) [9]. However, if the second-order reaction in eq. (12.13) is the rds, the peak or half-wave potential determined by a voltammetric method (cf. Chapter 11) will vary in a predictable way with the concentration of A [9].

$$A \underset{k}{\rightleftharpoons} B + e^- \qquad E^{o'}_1 \qquad (12.9)$$
$$B \xrightarrow{k} C \qquad\qquad (\text{rds}) \qquad (12.10)$$
$$B + C \rightleftharpoons A + D \qquad \text{fast}. \qquad (12.13)$$

Box 12.4 DISP1 pathway, reversible electron transfer.

Thus, systems with second-order rate-determining steps are easily recognized in preliminary voltammetric studies. For a first-order rds, peak potential does not depend on concentration of A.

In this section we consider only the ECE and DISP1 models, in which the pseudo-first-order conversion of B to C is the rds. We then discuss distinguishing between these two pathways for the hydroxylation of an organic cation radical.

The ECE model is given in Table 12.7. As for determination of rate constants with normal pulse voltammetry, the pulse width (τ) is critical for accurate estimation of k. Pointwise variance analysis showed that range $0 \le \log k\tau \le 1.7$ gave the smallest error in k. The use of τ values giving $\log k\tau$ well outside this range could give errors as large as 200% [8]. Considering that modern commercial instrumentation can easily provide 20–40 data points for $\tau = 20$ ms, the upper limit for determination of k by this method is about 2.5×10^3 s^{-1}.

Table 12.8 presents the model for the DISP1 pathway (Box 12.4). The same consideration applies as for the ECE mechanism, that the pulsewidth should obey the relation $0 \le \log k\tau \le 1.7$.

The models in Tables 12.7 and 12.8 were employed in a chronocoulometric study of the electrochemical hydroxylation of tetrahydrocarbazoles **1** and **2** [8], which are model compounds for structurally related anticancer indole alkaloids. Their anodic hydroxylation products in acetonitrile/water

Table 12.7 Model for ECE Pathway with Reversible Electron Transfer in Single Potential Step Chronocoulometry [8]

Assumptions: Reversible electron transfer, linear and edge diffusion; $D_O = D_R$; pseudo-first-order rds (Box 12.3)

Regression equation:

$$Q(t) = b_0 + b_1[4t^{1/2}\pi^{-1/2} - b_2{}^{-1/2}\text{erf}(b_2 t)^{1/2}] + b_3 t$$

$$b_0 = Q_{dl}; \; b_1 = FAD^{1/2}C;$$
$$b_2 = k; \; b_3 = aFADC/r$$

Q_{dl} = double layer charge; F = Faraday's constant, A = electrode area; D = diffusion coefficient; C = concentration of reactant; a = constant accounting for nonlinear diffusion; and r = radius of disk electrode

Regression parameters:	*Data:*
$b_0 \quad b_1 \quad b_2 \quad b_3$	$Q(t)$ vs. t

Special instructions: Use simplex or steepest decent algorithms for fitting data (Marquardt algorithm gave convergence problems) [10]. Correct range of pulse widths must be used for accurate k estimates [8]; that is, $-0 \le \log k\tau \le 1.7$. For some solid electrodes, background may need subtraction or consideration in the model. Compute erf from

$$(\pi^{1/2}/2)[\text{erf}(x)]/x = \exp(-x^2){}_1F_1(1, 2/3; x^2)$$

where

$${}_1F_1(a, b; x^2) = 1 + (^a/_b)z + a(a + 1)/[b(b + 1)]z^2/2! + \ldots$$

Kummer's confluent hypergeometric function, or from alternative standard series expansion [5].

Table 12.8 Model for DISPI Pathway with Reversible Electron Transfer in Single
Potential Step Chronocoulometry [8]

Assumptions: Reversible electron transfer, linear and edge diffusion; $D_O = D_R$; pseudo-first-
order rds (Box 12.4)
Regression equation:

$$Q(t) = b_0 + b_1\pi^{-1/2}[4t^{1/2} - 0.667b_2t^{3/2} - t^{1/2}\exp(b_2t) - 0.886\,\mathrm{erf}\{(b_2t)\}^{1/2}] + b_3t$$

$$b_0 = Q_{dl}; \qquad b_1 = FAD^{1/2}C;$$
$$b_2 = 2k; \qquad b_3 = aFADC/r$$

Q_{dl} = double layer charge; F = Faraday's constant; A = electrode area; D = diffusion
coefficient; C = concentration of reactant; a = constant accounting for nonlinear diffusion;
and r = radius of disk electrode

Regression parameters: *Data:*
   b_0 b_1 b_2 b_3 $Q(t)$ vs. t

Special instructions: Use simplex or steepest decent algorithms for fitting data (Marquardt
algorithm gave convergence problems) [10]. Correct range of pulsewidths must be used
for accurate k estimates [8]; that is, $-0 \le \log k\tau \le 1.7$. For some solid electrodes, background
may need subtraction or consideration in model. Compute erf from

$$(\pi^{1/2}/2)[\mathrm{erf}(x)]/x = \exp(-x^2)_1F_1(1, 2/3; x^2)$$

where

$$_1F_1(a, b; x^2) = 1 + (^a/_b)z + a(a + 1)/[b(b + 1)z^2/2! + \ldots$$

Kummer's confluent hypergeometric function, or from alternative standard series
expansion [5].

($>2M$) [11] are **3** and **4**. Carbon paste electrodes were used, which did not
require background subtraction.

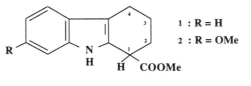

1 : R = H
2 : R = OMe

1-carboxymethyltetrahydrocarbazoles

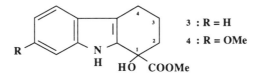

3 : R = H
4 : R = OMe

1-hydroxy-1-carboxymethyltetrahydrocarbazoles
(products)

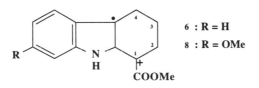

6 : R = H
8 : R = OMe

possible reactive intermediates in hydroxylation

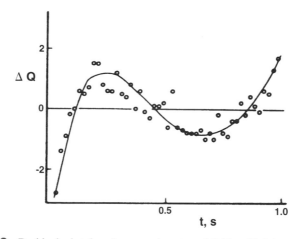

Figure 12.10 Residual plot for chronocoulometry of 0.65 mM **2** in acetonitrile with 11.1 M water and 0.2 M LiCLO$_4$ fit onto DISP1 model in Table 12.8. The solid line shows the residual (deviation) pattern for theoretical ECE data with $k = 25 \, s^{-1}$ fit onto DISP1 model in Table 12.8. (Reprinted with permission from [8], copyright by the American Chemical Society.)

Single potential step Q–t data for **1** and **2** obtained in acetonitrile containing 3–17 M water were analyzed by using the models in Tables 12.7 and 12.8. Comparisons of residual plots (cf. Section 3.C.1) were used to find the model that best fit the data. A typical residual plot from a fit of data onto the DISP1 model (Figure 12.10) showed a distinct nonrandom pattern, the same as for regression of theoretical noise-free ECE data onto the DISP1 model. When the same data were analyzed with the ECE model, random deviation plots were obtained (Figure 12.11). The data for both

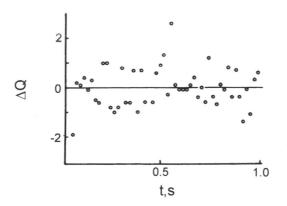

Figure 12.11 Residual plot for Q–t data used in Figure 12.10 analyzed by using the ECE model in Table 12.7. (Reprinted with permission from [8], copyright by the American Chemical Society.)

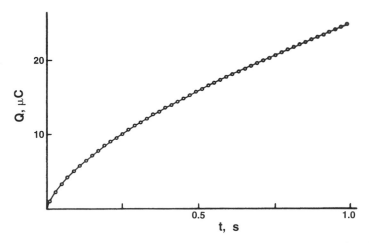

Figure 12.12 Experimental (○) and best fit calculated line (ECE model, Table 12.7) for chronocoulometric data used to obtain Figure 12.11. (Reprinted with permission from [8], copyright by the American Chemical Society.)

compounds studied were in excellent agreement with this model (Figure 12.12). Thus, the ECE model in Box 12.3 appears to be the best choice for the oxidation pathway of **1** and **2** [8].

The values of k_2 computed for **1** and **2** and their deuterio derivatives increased with decreasing concentration of water. The value of $\ln k_2$ of a reaction of an ion with a dipolar molecule in solution should increase approximately linearly with the inverse of the dielectric constant of the solution. This relation was obeyed rather well (Figure 12.13). Therefore, the differences in k_2 at different water concentrations were explained by the influence of water on the dielectric constant of the medium.

The results of these kinetic analyses were combined with quantum mechanical studies of possible reactants and intermediates to propose a detailed mechanism for the hydroxylations [8]. The proposed reactive intermediates are the cation radicals **6** and **8.**

D.3. Orthogonalized Model for the ECE Pathway

Readers should note the cautionary statements concerning the Marquardt–Levenberg algorithm in Tables 12.7 and 12.8. We have discussed the reasons for poor convergence of the ECE model with this algorithm in Section 4.A.2. Actually, the partial correlation of parameters in the models causes long convergence times even when using programs based on steepest descent or simplex algorithms, although the errors in the final parameters are negligible.

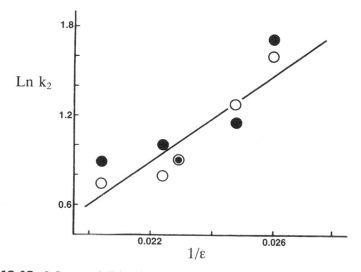

Figure 12.13 Influence of dielectric constant of the medium (ε) on k_2 for compounds **2** (○) and its **1**-deuterio derivative (●). (Reprinted with permission from [8], copyright by the American Chemical Society.)

Orthogonalization of the ECE model can be used to achieve better convergence properties when k is the main parameter of interest. The orthogonalization procedure and the ECE model is discussed in detail in Section 4.A.3. The orthogonalized model for the ECE pathway was used to analyze chronocoulometric data to obtain lifetimes as $t_{1/2} = 1/k$ for the spontaneous decomposition of aryl halide radicals [10].

References

1. A. J. Bard and L. R. Faulkner, *Electrochemical Methods*. New York: Wiley, 1980.
2. J. F. Rusling and Margaret Y. Brooks, "Analysis of Chronocoulometric Data and Determination of Surface Concentrations," *Anal. Chem.* **56** (1984), pp. 2147–2153.
3. A. Sucheta and J. F. Rusling, "Effect of Background Charge in Estimating Diffusion Coefficients by Chronocoulometry at Glassy Carbon Electrodes," *Electroanalysis* **3** (1991), pp. 735–739.
4. J. F. Rusling, "Analysis of Chemical Data by Computer Modeling," *CRC Critical Reviews in Analytical Chemistry* **21** (1989), pp. 49–81.
5. S. S. Shukla and J. F. Rusling, "Analyzing Chemical Data with Computers: Errors and Pitfalls," *Anal. Chem.* **56** (1984), pp. 1347A–1353A.
6. F. C. Anson, J. H. Christie, and R. A. Osteryoung, "A Study of the Adsorption of Cadmium(II) on Mercury from Thiocyanate Solutions by Double-Potential Step Chronocoulometry," *J. Electroanal. Chem.* **13** (1967), pp. 343–352.
7. M. K. Hanafey, R. L. Scott, T. H. Ridgeway, and C. N. Reilley, "Analysis of Electrochemical Mechanisms by Finite Difference Simulation and Simplex Fitting of Double-Potential

    Step Current, Charge, and Absorbance Responses," *Anal. Chem.* **50** (1978), pp. 116–137 and references therein.

8. J. F. Rusling, Margaret Y. Brooks, B. J. Scheer, T. T. Chou, S. S. Shukla, and Miriam Rossi, "Analysis of ECE-Type Electrode Reactions by Chronocoulometry: Anodic Hydroxylation of Tetrahydrocarbazoles," *Anal. Chem.* **58** (1986), pp. 1942–1947.

9. C. Amatore, M. Gareil, and J.-M. Saveant, "Homogeneous vs. Heterogeneous Electron Transfer in Electrochemical Reactions," *J. Electroanal. Chem.* **147** (1983), pp. 1–38 and references therein.

10. A. Sucheta and J. F. Rusling, "Removal of Parameter Correlation in Nonlinear Regression. Lifetimes of Aryl Halide Anion Radicals from Potential-Step Chronocoulometry," *J. Phys. Chem.* **93** (1989), pp. 5796–5802.

11. J. F. Rusling, B. J. Scheer, Azita Owlia, T. T. Chou, and J. M. Bobbitt, "Anodic Hydroxylation of 1-Carbomethoxy-1,2,3,4-tetrahydrocarbazoles," *J. Electroanal. Chem.* **178** (1984), pp. 129–142.

Chapter 13

Automated Resolution of Multiexponential Decay Data

A. Considerations for Analyses of Overlapped Signals

In Chapters 2 and 3, exponential decay of a signal was used extensively to illustrate various aspects of nonlinear regression analysis. Analysis of this type of data is quite important for processes such as luminescence decay, radioactive decay, and irreversible first-order chemical reactions. The general reaction is

$$A \xrightarrow{k_1} B. \tag{13.1}$$

Recall that the model for a single exponential decay of a signal y as a function of independent variable x (such as time) can be expressed as (Chapter 3)

$$y_j = b_1 \exp(-b_2 x_j) + b_3. \tag{13.2}$$

The parameters are b_1, the pre-exponential factor, $b_2 = k_1$ the rate constant for the decay, and b_3, a constant offset background. The decay lifetime is $\tau = 1/k_1$, or $1/b_2$.

Often, data containing two or more time constants are encountered. For mixtures of species that are all undergoing first-order decay with lifetimes τ_l, the appropriate model composed of sums of exponentials is given in Table 13.1. This model requires a judgment concerning the correct number of exponentials that best fit the data. The problem is complicated by the

Table 13.1 Model for Multiple Exponential Decay Kinetics

Assumptions: Background and extraneous instrumental signals unrelated to the decay have been removed from the data

Regression equation:

$$y = \sum_{i=1}^{p} y_i \exp\{-t/\tau_i\}; \qquad i = 1, \ldots, p$$

Regression parameters: *Data:*
  y_i and τ_i for $i = 1$ to p Signal (y) vs. time

Special instructions: Test for goodness of fit by comparing residual plots and by using the extra sum of squares F test (see Section 3.C.1); use proper weighting, such as for a counting detector $w_j = 1/y_j$ in eq. (2.9) (cf. Section 2.B.5)

---

fact that each successive higher order model contains two more parameters than its predecessor. That is, the single exponential contains two parameters, and a system with two decaying species contains four parameters. Thus, the best model can be chosen relying on comparisons of deviation plots and the extra sum of squares F test. As discussed in Section 3.C.1, the extra sum of squares F test provides a statistical probability of best fit that corrects for different numbers of parameters in two models.

Resolution of multiple exponentials becomes difficult when three or more exponentials overlap. Success in separating these contributions is a complex function of the relative values of y_i and τ_i. Degradation in the precision of these parameters for components with equal y_i values was found [1] when ratios of lifetimes were decreased from 5 to 2.3. Studies with simulated data showed similar results, although the difference between single, double, and triple exponentials could be ascertained [2] from noisy theoretical data for ratios of τ_2/τ_1 of 2.3 and τ_3/τ_2 of 4.3. However, for the triple exponential, the errors were 36% in τ_1, 23% in τ_2, and 7% in τ_3 [2]. As an approximate rule of thumb, nonlinear regression is not expected to give reliable results for τ values with the model in Table 10.1 for more than three overlapped exponentials ($p > 3$) or when the ratio of successive lifetimes falls below about 2.5.

These statements should be taken only as guidelines and need to be evaluated within the context of the problem to be solved. The success of the data analysis will also depend on the quality and the resolution of the data. For example, if an experiment produced a signal to noise (S/N) ratio of >1000, a data set containing >5000 individual points, and if all ratios of τ_{k+1}/τ_k were >5, it might be possible to fit these data with four exponentials. On the other hand, for data with (S/N) < 50, $\tau_{k+1}/\tau_k < 3$, and 30 individual data points, it might be difficult to get acceptable results even for a two or three exponential fit.

B. Automated Analysis of Data with an Unknown Number of Exponentials

A frequently encountered problem in luminescence or radioactive decay experiments is that the number of decaying species in a sample is not known. However, the number of decays can be found by a computer program. An automated method for analyzing exponential decay data with up to three contributing components was developed [2] based on deviation pattern recognition.

Deviation pattern recognition was developed by Meites [3] as a basis for decisions concerning the best model for a set of data. The technique employs the shapes of residual or deviation plots obtained after a regression analysis (see Chapters 2 and 3) to make automated decisions about goodness of fit. Recall that residual plots of dev_j/SD vs. x_j will have a random distribution of points around $dev_j/SD = 0$ if the regression model fits the data within the standard error in measuring the signal, y. If the model does not describe the data correctly, the residual plot will have a distinct nonrandom pattern. Scatter will be superimposed on this pattern reflecting the random noise in the signal.

Deviation patterns for fits onto a given model can be generated by nonlinear regression analysis of noise-free theoretical data sets representing other models in a library of models [3]. Information about these can be coded into the computer and used as a basis for construction the basis of a "decision tree" on which to base decisions in an expert system [4].

The multiexponential decay problem does not require the full power of deviation pattern recognition, because it is a linear classification problem. That is, we can test models with one, two, and three exponentials in sequence and accept the first model that gives an acceptable fit to the data based on deviation plots and other statistical criteria (Section C.3.1).

A partial program listing for automated analysis of exponential decay data is given in Box 13.1. This program [2] assumes that the data were collected by using a counting detector, and the regression analysis is therefore weighted for Poisson statistics (Section 2.B.5). It also assumes that all extraneous signals, such as a source light pulse for example, have been removed prior to the analysis of the data. The computer program sequentially fits models to the experimental data involving one, two, and three exponentials, stopping when an acceptable model is found. The choice of the correct model is based on achieving predefined criteria of randomness in smoothed residual plots.

The program in Box 13.1 estimates initial values of y_i and τ_i by preliminary *linear* regressions. At the beginning of the program, a linear regression of all data as $\ln y$ vs. t are used to estimate initial values of y_1 and τ_1, and these values are employed to begin the *nonlinear* regression analysis onto the model in Table 13.1 with $p = 1$.

Box 13.1 FORTRAN code for automated analysis of exponential decay data. (This listing should be modified for linkage to an appropriate nonlinear regression subroutine. The nonlinear regression program referred to in the listing is denoted NONLINEAR REGRESSION.)

```
PROGRAM FOR AUTOMATIC DECONVOLUTION OF EXPONENTIAL DECAYS

LINE(S)   STATEMENT

1             IMPLICIT INTEGER (T)
2             INTEGER P,PLOT
3             LOGICAL*1 HEAD(80)
4             DOUBLE PRECISION X,Y,Q,V,W,Y1,U,G,D,DO,DABS,DLOG,
5             LDSIGN,DFLOAT,DEXP,DSQRT,C,SS,S1,M,S,SS2,AOUT1,AOUT3,
6             LR,DARSIN,DIFF,ZX,S9
7             COMMON X(100),Y(100),Q(100),V(12),W(100),Y1(100),
8             LPT(100),U(12),DIFF(100),S,IO,N1,T,T1,Q1,JA,DO,J8,IWT,I9,J9
9             DATA LINP/5/,LOUT/6/
10     1      READ (LINP,10,END=15) HEAD
11    10      FORMAT(80A1)
12  9100      FORMAT(3I1)
13  9200      FORMAT(2I3,I4,D6.4)
14            READ(LINP,9100) T,T1,PLOT
15            READ(LINP,9200) P,IO,N1,DO
16            READ(LINP,*) IWT
17            J9=0
18            JA=0
19            DO 70 N=1,N1
20    70      READ(LINP,*) Y(N)
21            DO 90 N=10,64
22    90      X(N-9)=N*20. + 10.
23            CALL GRAPH
24    75      CALL LINREG
25            CALL NONLINEAR REGRESSION(P,DO,HEAD,PLOT)
       (      SUBROUTINE FOR NONLINEAR REGRESSION)
26            GO TO 74
27    74      DO 76 N=1,N1
28            X(N)=X(N)
29            IF(N)=Y(N)
30    76      CONTINUE
31            IO=IO + 2
32            IF(IO-7) 79,79,15
33    79       DO 80 I=1,IO
34            D(I)=DABS(V(I))/2
35    80      CONTINUE
36            J9=J9 + 1
37            GO TO 75
```

```
38    15    WRITE(LOUT,200)
39          CALL EXIT
40    200   FORMAT('O+++++++++PROGRAM '1'TERMINATION++++++++'/
41    1      'O END OF FILE ENCOUNTERED DURING DATA READ')
42          STOP
43          END
44          SUBROUTINE GRAPH
45          LOGICAL *1 LINE(65),BLANK,STAR,CHRS(10),HEAD(80)
46          INTEGER N,PT,J,IDIG
47          DOUBLE PRECISION X,Y,Q,V,W,Y1,U,G,D,D0,DABS,DLOG,
48          LDSIGN,DFLOAT,DEXP,DSQRT,C,SS,S1,M,S,SS2,AOUT1,AOUT3,
49          LR,DARSIN,DIFF,ZX,S9
50          COMMON X(100),Y(100),Q(100),V(12),W(100),Y1(100),
51          LPT(100),U(12),DIFF(100),S,IO,N1,T,T1,Q1,JA,DO,J8,IWT,I9,J9
52          COMMON/COMDEV/AOUT2(100)
53          DATA CHRS/'0','1','2','3','4','5 ','6','7','8','9',
54          L'*'/
55          DATA BLANK/' '/,STAR/'*'/
56          DATA LINP/5/,LOUT/6/
57          WRITE(LOUT,2)
58    2     FORMAT(5X,22HGRAPH OF LN(I) VS TIME)
59          DO 1 N=1,N1
60    1     AOUT2(N)=DLOG(Y(N))
61          I8=60/INT((AOUT2(I)-AOUT2(N1))*10.)
62          N3=N1 + 1
63          DO 840 N2=1,N3
64          DO 810 J=1,65
65    810   LINE(J)-BLANK
66          N=N1-N2 + 1
67          IF(N) 827,827,812
68    812   PT(N)-INT(AOUT2(N)*10)
69          LINE(62)=STAR
70          I9=INT(ABS(AOUT2(I)*10))*I8
71          I7=I9-60
72          IF (I9-IABS(PT(N))) 825,820,820
73    820   J=PT(N)*I8-I7
74          LINE(J)=STAR
75    825   IF (N-(N/10)*10) 830,826,830
76    826   LINE(62)-CHRS(N/10 + 1)
77          LINE(63)=CHRS(1)
78          GO TO 830
79    827   DO 829 J=1,62
80    829   LINE(J)=STAR
81    830   WRITE(LOUT,855) LINE
82    855   FORMAT(1H,6X,65A1)
```

Box 13.1 (*continues*)

```
83   840   CONTINUE
84         STOP
85         END
86         SUBROUTINE LINREG
87         DOUBLE PRECISION X,Y,Q,V,W,Y1,U,G,D,D0,DABS,DLOG,
88         LDSIGN,DFLOAT,DEXP,DSQRT,C,SS,S1,M,S,SS2,AOUT1,AOUT3,
89         LR,DARSIN,DIFF,ZX,S9
90         COMMON X(1OO),Y(1OO),Q(100),V(12),W(100),Y1(100),
91         LPT(100),U(12),DIFF(100),S,I0,N1,T,T1,Q1,JA,D0,J8,IWR,I9,J9
92    10   IF(J9) 13,13,21
33    21   IF(J9-1) 80,11,22
94    22   IF(J9-2) 80,14,23
95    23   IF(J9-3) 80,15,24
56    24   IF(J9-4) 80,11,25
97    25   IF(J9-5) 80,16,26
98    26   IF(J9-6) 80,17,27
99    27   IF(J9-7) 80,18,28
100   28   IF(J9-8) 80,11,11
101   13   N2=1.
102        N3=N1
103        GO TO 12
104   14   N2=1
105        N3=N1/2
106        GO TO 12
107   15   N2=N1/2
108        N3=N1
109        GO TO 12
110   16   N2=1
111        N3=N1/3
112        GO TO 12
113   17   N2=N1/3
114        N3=2*N1/3
115        GO TO 12
116   18   N2=2*N1/3
117        N3=N1
118        GO TO 12
119   11   RETURN
120   12   DO 20 I=N2,N3
121        Y1(I)=DLOG(Y(I))
122        W(I)=1.
123        IF(IWT.EQ.1) GO TO 20
124        W(I)=1./Y1(I)
125   20   CONTINUE
```

```
126            GO TO 70
127     70     SUM=O
128            SUMX=0.
129            SUMY=0
130            SUMXY=0.
131            SUMXX=0.
132            SUMYY=0.
133            DO 100 I=N2,N3
134            SUM=SUM+W(I)
135            SUMX=SUMX+X(I)*W(I)
136            SUMY=SUMY+Y1(I)*W(I)
137            SUMXY=SUMXY+X(I)*Y1(I)*W(I)
138            SUMXX=SUMXX+X(I)*X(I)*W(I)
139    100     SUMYY=SUMYY+Y1(I)*Y1(I)*W(I)
140            SDELT=SUM*SUMXX-SUMX*SUMX
141            SA=(SUMXX*SUMY-SUMX*SUMXY)/SDELT
142            SB=(SUM*SUMXY-SUMX*SUMY)/SDELT
143            SRHO=ABS((SUM*SUMXY-SUMX*SUMY)/
144           1SQRT(ABS(SDELT*(SUM*SUMYY-SUMY*SUMY)))))
145            SIGMA=(SUMYY-SA*SUMY-SB*SUMXY)/((N3-N2)-2.)
146            SIGMA=ABS(SIGMA/((N3-N2)-2.0))
147            STDA=SQRT(ABS(SIGMA*SUMXX/SDELT))
148            STDB=SQRT(ABS(SIGMA*SUM/SDELT))
149     41     IF(J9) 80,31,42
150     42     IF(J9-2) 80,32.43
151     43     IF(J9-3) 80,33,44
152     44     IF(J9-5) 80,34,45
153     45     IF(J9-6) 80,35,46
154     46     IF(J9-7) 80,36,36
155     31     V(1)=EXP(SA)
156            V(2)=-1./SB
157            GO TO 50
158     32     V(1)=EXP(SA)
159            V(2)=-1./SB
160            GO TO 50
161     33     V(3)=EXP(SA)
162            V(4)=-1./SB
163            GO TO 50
164     34     V(1)=EXP(SA)
165            V(2)=-1./SB
166            GO TO 50
167     35     V(3)=EXP(SA)
168            V(4)=-1./SB
169            GO TO 50
170     36     V(5)=EXP(SA)
171            V(6)--1./SB
```

Box 13.1 (*continues*)

```
172          GO TO 50
173    50    IF(IWT.EQ.0)    WRITE(6,57)
174          IF(IRT.EQ.1)    WRITE(6,58)
175          WRITE(6,66)V(1),STDA,V(2),STDB,SIGMA,SRHO
176          J9=J9+1.
177          GO TO 10
178    80    STOP
179    57    FORMAT(//'       STATISTICAL WEIGHTING, W(I)=1./Y(I)')
180    58    FORMAT(//'       UNIT WEIGHTING, W(I)=1.0')
181    B6    FORMAT(//' FOR Y=A + BX, A= ',1PD12.3,
182    1     ' +/- '1PD10.2,//21X, B=',1PD12.3,' +/- ,1PD10.2.
183    2     //' THE VARIANCE OF THE FIT IS',1PD12.3,
184    3     //' AN.D THE CORRELATION COEFFICIENT IS ',1PD12.3//}
185    1     ' STARTING OVER.'/)
186          END
187          SUBROUTINE NONLINEAR REGRESSION(P,D0,HEAD,PLOT)
352    456   AOUT2(N)=AOUT1/(SS2*DSQRT(Y(N)))
360          DO 500 I=1,N5
361          SUM=O
362          SUM=SUM + AOUT2(I) + AOUT2(I+1) + AOUT2(I+2)
363          SUM=ABS(SUM)
364          WRITE(6,1461) SUM
365          IF(SUM-3.75) 500,500,490
             (THRESHOLD TEST - THRESHOLD MAY BE CHANGED BY THE USER)
366    1461  FORMAT(5X,1PD13.6)
367    500   CONTINUE
368          JA=JA+1
423          SUBROUTINE ERRSUM(II,*)
             (USE THE FOLLOWING MODIFICATION FOR POISSON WEIGHTING)
437    521   S=S+((Y(N)-Q(N))**2)*(1./Y(N))
44Y          SUBROUTINE PLOT1(HEAD)
(A SUBROUTINE IS REQUIRED HERE TO MAKE A GRAPH OF THE DEVIATION PLOT)
```

```
490         SUBROUTINE CALC -
            (TO BE INSERTED INTO THE NONLINEAR REGRESSION SUBROUTINE,
            V(K) ARE THE MODEL PARAMETERS)
491         DOUBLE PRECISION X,Y,Q,V,W,Y1,U,G,D,D0,DABS,DLOG,
492         LDSIGN, DFLOAT, DEXP, DSQRT,C,SS,S1,M,S.SS2,AOUT1,AOUT3,
493         LR,DARSIN,DIFF,ZX,S9
494         COMMON X(100),Y(100),Q(100),V(12),W(100),Y1(100),
495         LPT(100),U(12),DIFF(100),S,I0,N1,T,T1,Q1,JA,DO,J8,IWT,I9,J9
496         IF(I0-3) 24,24,27
497    24   IF(V(2)-10.) 3,3,25
498    3    DO 4 N=1,N1
499    4    Q(N)=Q(N)*1.01
500    25   DO 26 N=1,N1
501    26   Q(N)=V(1)*DEXP(-X(N)/V(2))
502         RETURN
503    27   IF(I0-5) 29,29,33
504    29   IF(V(2)-10.) 6,6,5
505    5    IF(V(3)-.001) 6,6,8
506    8    IF(V(4)-10.) 6,6,30
507    6    DO 7 N=1,N1
508    7    Q(N)=Q(N)*1.01
509    30   DO 31 N=1,N1
510    31   Q(N)=V(1)*DEXP(-X(N)/V(2)) + V(3)*DEXP(-X(N)/V(4))
511         RETURN
512    33   IF(I0-7) 37,37,36
513    37   IF(V(2)-10.) 9,9,12
514    9    V(2)=DABS(V(2)*50.)
515    12   IF(V(4)-10.) 10,10,13
516    10   V(4)=DABS(V(4)*50.)
517    13   IF(V(6)-10.) 11,11,34
518    11   V(6)=DABS(V(6)*50.)
519    34   DO 35 N=1,N1
520    35   Q(N)=V(1)*DEXP(-X(N)/V(2)) + V(3)*DEXP(-X(N)/V(4)) +
521         LV(5)*DEXP(-X(N)/V(6))
522         RETURN
523    36   STOP
524         END
```

Note: that Subroutine CALC is to be inserted into the nonlinear regression
subroutine used with this listing

Box 13.1 (*continues*)

DOCUMENTATION:

LINE(S)	COMMENTS
1	MAIN PROGRAM

2 - 9   DEFINE VARIABLES IN DOUBLE PRECISION, COMMON
STATEMENT, ETC.

10   READ HEADING FOR DEVIATION PLOT.

11 - 13   FORMAT STATEMENTS FOR HEADING, T, T1, PLOT, AND P,
I0, N1, D0.

14   READ T, T1, PLOT. T=0 (ABSOLUTE ERRORS) OR T=1
(RELATIVE ERRORS) T1=1 (SIGNS CERTAIN) T1=0 (SIGNS

UNCERTAIN P=1 (DEVIATION PLOT DESIRED) P=0 (NO
DEVIATION PLOT).

15   READ P, I0, N1, D0. P=NUMBER OF CYCLES BETWEEN
PRINTOUT OF THE ERRORSUM; I0 REPRESENTS THE NUMBER
OF PARAMETERS; N1 THE TOTAL NUMBER OF DATA POINTS;
D0 THE ESTIMATED RELATIVE ERROR IN THE PARAMETERS.

16   READ IWT, A WEIGHTING FACTOR; 1=UNIT WEIGHTING,
0=STATISTICAL WEIGHTING.

17   READ J9, THE COUNTER THAT INDICATES WHICH SECTION OF
DATA TO USE FOR LINEAR REGRESSION. I.E. FIRST 1/3,
SECOND 1/3, OR THIRD 1/3 OF DATA TO DETERMINE THE
PARAMETER VALUES V(1), V(2); V(3), V(4); AND V(5),
V(6), FOR A THREE EXPONENTIAL CURVEFIT.

18   READ JA, THE COUNTER FOR THE NUMBER OF TIMES
NONLINEAR REGRESSION HAS BEEN PERFORMED.

19 - 22   READ Y DATA. CALCULATE X DATA ACCORDING TO
CHANNEL NUMBER AND THE TIME EACH CHANNEL REPRESENTS.
EXAMPLE: IF EACH CHANNEL REPRESENTS 10 MSEC AND
CHANNELS 90 TO 190 ARE BEING READ, THE COMPUTER
STATEMENT SHOULD READ:
      DO 90 N=90,190
   90 X(N-89)=N*10.

23   CALL "GRAPH" TO GRAPH LOG_e(INTENSITY) VS. TIME.
THIS MAY BE AN AID TO DETERMINING HOW MANY

EXPONENTIAL DECAYS THE DATA CONTAINS.

24   CALL "LINREG" WHICH DOES A LINEAR REGRESSION OF
            LOG$_E$(INTENSITY) VS. TIME USING ALL THE DATA TO
            DETERMINE THE PRE-EXPONENTIAL FACTOR AND
            FLUORESCENCE LIFETIME (FOR 1 EXPONENTIAL DECAY); THE
            DATA IS DIVIDED IN HALF TO CALCULATE TWO PRE-
            EXPONENTIAL FACTORS AND TWO FLUORESCENCE LIFETIMES
            (FOR 2 EXPONENTIAL DECAY); THE DATA IS DIVIDED INTO
            THIRDS TO CALCULATE 3 PRE-EXPONENTIAL FACTORS AND 3
            FLUORESCENCE LIFETIMES (FOR 3 EXPONENTIAL DECAY).

25   AFTER DETERMINING ROUGH VALUES FOR PRE-EXPONENTIAL
            FACTORS AND FLUORESCENCE LIFETIMES, SEND THE
            INFORMATION TO (NONLINEAR REGRESSION
            SUBROUTINE).

26   AFTER CURVEFIT HAS BEEN COMPLETED, GO TO STATMENT
            74.

27 - 30   (STATEMENT 74) ESTABLISH VALUES FOR X(N), AND Y(N).

31   INCREASE BY TWO THE NUMBER OF PARAMETERS.

32   IF THE NUMBER OF PARAMETERS IS GREATER THAN 7, PRINT
            "PROGRAM TERMINATION", IF NOT, GO TO STATEMENT 79.

33 - 35   (STATEMENT 79) ESTABLISH NEW VALUES FOR RELATIVE
            ERROR IN THE PARAMETERS.

36   INCREMENT COUNTER FOR LINEAR REGRESSIONS

37   START FITTING SERIES AGAIN

38 - 43   STOP PROGRAM

44   SUBROUTINE GRAPH (GRAPHS LOG$_E$(INTENSITY) VS. TIME)

45   THESE ARE LOGICAL VARIABLES USED IN GRAPH SUBROUTINE

46   DEFINE INTEGERS

47 - 52   DEFINE DOUBLE PRECISION VARIABLES

53 & 54   DEFINE CHARACTERS USED IN GRAPH SUBROUTINE

Box 13.1 (*continues*)

55   DEFINE BLANK AND STAR

56   LINP=5, LOUT=6

57 & 58   WRITE HEADING

59 & 60   DEFINE AOUT2(N) = LOG_E(INTENSITY)

61   CALCULATE I8, THAT IS USED TO SCALE THE GRAPH.

62   CALCULATE N3 = TOTAL NUMBER OF LINES IN THE GRAPH

63   START THE LOOP TO PRINT THE GRAPH.

64 & 65   LEAVE LINE BLANK.

66   CALCULATE A VALUE FOR N, THE DATA POINT NUMBER.

67   IF N IS LESS THAN 0 GO TO STATEMENT 827 WHICH SAYS
TO PRINT A LINE OF ASTARISKS. OF N IS GREATER THAN
0 GO TO STATEMENT 812.

68   MAKE AN INTEGER OUT OF LOG_E(INTENSITY)*10.=PT(N)

69   POSITION AN ASTARISK IN THE RIGHT HAND MARGIN OF
THE GRAPH.

70 - 73   SCALE THE DATA TO FIT THE GRAPH.

74   POSITION THE ASTARISK WHERE THE DATA POINT LIES ON
THE GRAPH.

75 - 77   PRINT THE DATA POINT NUMBER IN THE MARGIN EVERY
10TH POINT.

78   GO TO 830 WHICH SAYS TO PRINT "LINE". LINE CONSISTS
OF AN ASTARISK AT THE CORRECT POSITION FOR THE DATA
POINT VALUE AND AN ASTARISK (OR NUMBER) IN THE
MARGIN.

79 & 80   IF "LINE" IS THE LAST LINE, PRINT A LINE OF
ASTARISKS AT THE END OF THE GRAPH.

81 & 82   (STATEMENT 830) PRINT "LINE".

83   CONTINUE PROCEDURE TO CALCULATE AND PRINT ANOTHER

LINE.

84 & 85   END PROGRAM

86   SUBROUTINE LINREG (DETERMINES PRE-EXPONENTIAL FACTORS AND FLUORESCENCE LIFETIMES)

87 - 91   DOUBLE PRECISION VARIABLES AND COMMON STATEMENT.

92 - 119   THESE STATEMENTS DIRECT THE PROGRAM TO DIVIDE THE DATA ACCORDING TO THE NUMBER OF NONLINEAR REGRESSIONS WHICH HAVE BEEN DONE.

120 - 126   DETERMINE THE WEIGHTING FACTORS.

149 - 172   CALCULATE VALUES FOR PRE-EXPONENTIAL FACTORS AND FLUORESCENCE LIFETIMES.

173 - 186   OUTPUT FOR THE SUBROUTINE. GIVE THE TYPE OF WEIGHTING, INCREMENT J9 - THE COUNTER WHICH INDICATES WHICH SECTION OF DATA TO USE.

187   SUBROUTINE CFT4A - STATEMENTS ARE THE SAME AS THEY ARE IN THE ORIGINAL EXCEPT FOR THE LINES INDICATED BELOW.

352   AOUT2(N) = (Q(N)-Y(N))/SS2*SQRT(Y(N)) WHERE SS2 IS THE STANDARD DEVIATION OF REGRESSION. DIVISION BY SQRT(Y(N)) IS FOR STATISTICAL WEIGHTING.

360 - 368   SUM UP 3 SUCCESSIVE DEVIATIONS AND COMPARE THE SUM TO 4. IF THE SUM > 4, THIS INDICATES THAT THE PLOT IS NONRANDOM AND THE PROGRAM GOES TO STATEMENT 490 (WHICH INCREMENTS COUNTER JA, EQUAL TO THE NUMBER OF CURVEFITS, BY 1 AND RETURNS TO THE MAIN PROGRAM TO START THE NEXT CURVEFIT). IF THE SUM < 4, INDICATING THAT THE DEVIATION PLOT IS RANDOM, THE PROGRAM GOES TO STATEMENT 500 (WHICH INCREMENTS COUNTER JA, BY 1 AND CONTINUES ON TO PRINT OUT THE STATISTICS AND THE DEVIATION PLOT).

423   SUBROUTINE ERRSUM

Box 13.1 (*continues*)

done separately on each half-set of the data to obtain the initial guesses of y_i and τ_i. Nonlinear regression analysis onto the $p = 2$ model is then begun using these initial parameters. After convergence, the threshold testing of the new smoothed deviation plot is done as described. If no new D_s exceeds the threshold, the $p = 2$ hypothesis is accepted. If one or more D_s exceed the threshold, a third regression analysis with $p = 3$ is initiated.

Initial parameter guesses for this final nonlinear regression onto the $p = 3$ model are chosen by three semilogarithmic linear regressions of the data divided into three equal parts. After the regression analysis, the new smoothed deviation plot is analyzed as previously to test the $p = 3$ hypothesis.

The automatic classification program was tested with data simulated from the model in Table 13.1, with 1% Poisson distributed random noise added. Attention was paid to the success of classifications having relatively small ratios of lifetimes; that is, τ_{i+1}/τ_i. Results for simulated data with $\tau_2/\tau_1 = 4.3$ and $\tau_3/\tau_2 = 4.3$, $\tau_2/\tau_1 = 2.3$ all gave successful classification. Because their values were relatively close, errors for τ_2 and τ_1 were in excess of 20%. Errors deceased significantly as the ratio τ_2/τ_1 in the simulated data was increased [2].

Table 13.2 also lists the initial estimates of the parameters found by the program using linear regression. For the single exponential ($p = 1$), these values were quite similar to the final values obtained by nonlinear regression. However, for the data representing two and three exponentials, the final parameters from nonlinear regression had significantly smaller errors.

Table 13.2 Automated Classification and Analysis of Simulated Data[a]

p	Parameter	True value	Initial value[b]	Initial % error	Final value	Final % error
1	y_1	1878	1888	0.5	1888	0.5
	τ_1	698	695	0.4	694	0.5
2	y_1	9080	6738	26	8762	3.5
	τ_1	162	319	97	170	5.0
	y_2	1878	2441	30	1723	8.2
	τ_2	698	609	12	732	4.9
3	y_1	20771	10219	51	15932	23
	τ_1	69	245	255	94	36
	y_2	9080	3800	58	5265	42
	τ_2	162	457	180	199	23
	y_3	1878	2354	25	1638	13
	τ_3	698	621	11	747	7

[a] Using program listed in Box 13.1.
[b] Initial parameters from linear regression.

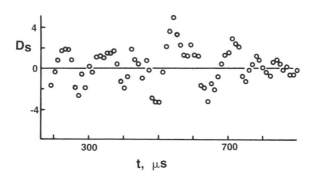

Figure 13.2 Smoothed, weighted deviation plot for luminescence decay of uranyl-Na mordenite regressed onto model for triple exponential decay using program in Box 10.1. (Reprinted with permission from [2], copyright © 1985, Marcel Dekker, Inc.)

The automatic classification method was applied to luminescence decay data for different types of zeolites (molecular sieves) containing uranyl ions [2]. Multiple exponential decays are observed corresponding to different environments for uranyl ions in the zeolite [5]. When data for uranyl in a sodium mordenite zeolite were analyzed by the program in Box 13.1, the smoothed deviation plot from the $p = 1$ model was clearly nonrandom (Figure 13.1), and it was classified as such by the threshold analysis employed by the program. The $p = 2$ hypothesis was also rejected by the program. The subsequent $p = 3$ fit gave a smoothed deviation plot (Figure 13.2) with no points exceeding the threshold, and the $p = 3$ hypothesis was accepted [6]. The same conclusion was reached independently by the inorganic chemists who collected the data after examining the results of individual nonlinear regressions onto one, two, and three exponential models [5]. Parameter values were in good agreement with those reported previously.

A problem in classification was noted if the data contained high frequency instrumental noise. As in most nonlinear regression applications, the method requires the best S/N possible and elimination of systematic noise. The program has the capability of analyzing only exponentials, so the data should contain no contributions from the source or from events with nonexponential responses.

References

1. J. N. Demas, *Exited State Lifetime Measurements.* New York: Academic Press, 1983.
2. M. Y. Brooks and J. F. Rusling, "Automatic Computerized Analysis of Multiexponential Decay Curves," *Anal. Lett.* **18** (1985), pp. 2021–2032.

3. L. Meites, "Some New Techniques for the Analysis and Interpretation of Chemical Data," *CRC Critical Reviews in Analytical Chemistry* **8** (1979), pp. 1–53.

4. J. F. Rusling, "Analysis of Chemical Data by Computer Modeling," *CRC Critical Reviews in Analytical Chemistry* **21** (1989), pp. 49–81.

5. S. L. Suib, A. Kostapapas, and D. Psaras, "Photoassisted Catalytic Oxidation of Isopropyl Alcohol by Uranyl-Exchanged Zeolites," *J. Amer. Chem. Soc.* **106** (1984), pp. 1614–1620.

6. M. Y. Brooks, "Applications of Computerized Analysis and Interpretation of Data in Chronocoulometry and Luminescence Decay," Ph.D. thesis, Storrs: University of Connecticut, 1984.

Chapter 14

Chromatography and Multichannel Detection Methods

In this chapter, we present specific examples of the use of nonlinear regression analysis to resolve signals in multicomponent analyses. We first discuss resolution of chromatographic peaks with single-channel detection, including examples from high-pressure liquid chromatography and high-performance size-exclusion chromatography. Following this, we discuss powerful and information-rich multichannel methods characteristic of chromatography coupled with detection by mass, infrared or UV-VIS detection. Also included is a section on multicomponent time-resolved fluorescence, which is representative of any time resolved spectroscopic method. Although this chapter could contain examples of many more techniques, we wish to stress the generality of the approach that can be applied to other similar types of methods.

A. Overlapped Chromatographic Peaks with Single-Channel Detection

A.1. Models for Peak Shape

This section discusses analysis of data from single channel chromatography. In these experiments, the effluent from the column passes through a single detector and the output of the analysis is the response of the detector vs. time. The time of introduction of the sample onto the column is $t = 0$. In Section 3.B.2 we discussed various models for peak-shaped data. Because of the wide use of chromatography in modern science, we now focus specifically on separation of overlapped peaks in this technique.

Full baseline resolution of all sample components is the ultimate goal of a chromatographic analysis, but it is not always achievable. This is especially true for complex samples containing many components with similar properties. If the shapes of chromatographic peaks are perfectly symmetrical, they can often be modeled using Gaussian or Gaussian–Lorentzian peak shapes. Models for overlapped peaks identical to those described in Section 3.B.2 can be employed in such cases.

The symmetric peak shape represents an ideal case in chromatography. Peaks frequently have some degree of asymmetry, featuring so-called tailing on the longer time side of the peak. Examples of symmetric and tailing peak shapes are given in Figure 14.1.

Research into models for unsymmetric chromatographic peaks has been conducted for several decades [1–3]. We will provide here only a brief summary of some of the more useful models considered. At the end of this section, we discuss simple, easy to compute, general models (Tables 14.1 and 14.2).

One appropriate and popular model for chromatographic peaks involves the convolution of a Gaussian peak with an exponential tail [1]. This model, known as the *exponentially modified Gaussian* [2] can be expressed as

$$R = \frac{1}{\tau} \int_{-\infty}^{t} dt' \exp[(t - t_0)^2/2W^2] \exp[-(t - t' + t_0)/\tau] \qquad (14.1)$$

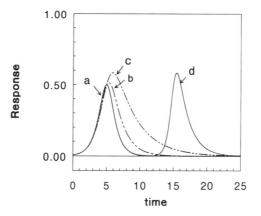

Figure 14.1 Examples of symmetric (a) and tailing (b–c) chromatographic peaks computed from the model in Table 14.1. Parameters used to compute the peaks are as follows:

Peak/parameter	A	c	a	b
a	1	5	1	1
b	1	5	1	0.6
c	1	5	1	0.3
d	1	15	2	0.6

where

  R = response of the detector,
  t = time following injection of the sample,
  t_0 = position of the peak maximum on the t axis,
  τ = decay lifetime of the tail,
  W = width at half-height of the Gaussian component of the peak.

The model in eq. (14.1) was found to be identical to a Gaussian model when $\tau/W < 0.4$. Approximate algorithms can be used to compute the model when $\tau/W > 0.4$, and these are described in the original literature [1]. A computer program for model computation has been published [2]. The form of eq. (14.2) relates the model specifically to chromatographic parameters, as follows:

$$R = A \frac{\sigma_G \sqrt{2}}{\tau} \exp\left[0.5\left(\frac{\sigma_G}{\tau}\right)^2 - \frac{t - t_G}{\tau} \right] \int_{-\infty}^{z/\sqrt{2}} \exp(-x^2)dx \quad (14.2)$$

where

$$z = \frac{t - t_G}{\sigma_G} - \frac{\sigma_G}{\tau}$$

and

  t_G = retention time,
  σ_G = standard deviation of the Gaussian function,
  A = constant proportional to peak amplitude,
  τ = time constant of the exponential decay.

Although the exponentially modified Gaussian model has been quite useful, its computation is somewhat cumbersome. Other models that have been suggested include two Gaussian peaks, each with different widths, a Poisson distribution, and a log normal distribution.

A closed-form model that is general and quite simple to compute was published [3] about the same time as the writing of this chapter. This model can provide both asymmetric (i.e., tailing) and symmetric peaks. It is summarized in Table 14.1. The symmetry of the peak depends on the relative values of parameters a and b. The height of the peak depends mainly on A, but also on a and b. Peak width depends on the magnitude of a and b. For example, the peak becomes less wide as the values of a and b increase when a/b is kept constant.

Figure 14.1 shows a series of peaks computed with the function in Table 14.1 for identical values of A. The solid line at $t = 5$ (a in Figure 14.1) illustrates the symmetric peak obtained when $a = b$. This curve is a hyperbolic secant, which has a shape roughly intermediate between Gaussian and Lorentzian [3]. When $b < a$, tailing is observed as shown in b–d of

Table 14.1 General Model for Symmetric and Unsymmetric Single Peaks [3]

Properties: $a > b > 0$ for a tailing peak; $a = b$ for a symmetric peak, A is related to peak height, and c controls the peak position; the term $b_5t + b_6$ is the background
Regression equation

$$R = \frac{A}{\exp[-a(t - c)] + \exp[-b(t - c)]} + b_5t + b_6$$

Regression parameters: *Data:*
   A a b c b_5 b_6 Response vs. time
Useful quantities:

$$\text{Peak position: } t_0 = c + \frac{\ln(a/b)}{a + b}$$

$$\text{Peak height: } h = \frac{(a)^{\frac{a}{a+b}}(b)^{\frac{b}{a+b}}}{a + b} A; \text{ between 0.5 and 1 when } A = 1$$

$$\text{Width at half-max: } W = \frac{(a + b)}{ab} \ln\left[\frac{1}{h} + \sqrt{\frac{1}{h^2} - \cos\left\{\frac{\pi(a - b)}{2(a + b)}\right\}}\right]$$

Figure 14.1. Curve d has the same a/b as c in Figure 14.1, but the values of a and b have been doubled and c shifted to a larger value for clarity. This change illustrates the decrease in peak width given by this function as absolute values of a and b are increased.

Overlapped peaks can be resolved by using a model consisting of the sum of two functions of the form in Table 14.1. This overlapped peak model (Table 14.2) provides rapid and stable convergence during nonlinear regression analysis using the Marquardt–Levenberg algorithm, provided that good initial guesses of peak positions are used. Initial guesses of the peak positions can be made using the minima of second derivatives, as discussed in Chapter 7. The peak position is relatively close to the parameter c_j (see Table 14.1). The A_j values are roughly twice the heights of the component peaks.

An example of a two-peak fit to theoretical data illustrates the quality of the initial guesses required for good convergence. Figure 14.2 shows a set of data for two overlapping peaks computed from the multiple peak model in Table 14.2 with normally distributed absolute error of $0.5\% A_1$. Also shown in this figure is the theoretical curve computed from the initial guesses used to begin a regression analysis, whose results are illustrated in Figure 14.3. Although the curve computed from initial guesses shows deviations of data from the model in all regions of the peak, the regression analysis converged to an excellent fit after six cycles. The parameters found in this analysis are given in Table 14.3.

These results illustrate how close our initial guesses need to be to achieve rapid convergence of the model in Table 14.2. Larger errors in initial values of A, a, and b caused excessively long convergence times using the

Table 14.2   General Model for Multiple Symmetric or Unsymmetric Single Peaks

Properties: $a > b > 0$ for a tailing peak; $a = b$ for a symmetric peak; A is related to peak height and c controls the peak position; the term $b_5t + b_6$ is the background
Regression equation:

$$R = \sum_{j=1}^{k} \left\{ \frac{A_j}{\exp[-a_j(t - c_j)] + \exp[-b_j(t - c_j)]} \right\} + b_5t + b_6$$

for k peaks

Regression parameters:					*Data:*
A_j	a_j	b_j	c_j	for $i = 1$ to k	Response vs. time
and	b_5	b_6			

Special instructions:

1. Second derivative of chromatogram can be used to identify number of peaks and their initial positions in cases of severe overlap (see Chapter 7)
2. Test for goodness of fit for models with different numbers of peaks by comparing residual plots and by using the extra sum of squares F test (see Section 3.C.1)
3. Peak properties can be obtained from formulae in Table 14.1; peak areas can be obtained by integration of the component peaks after finding their parameters in the regression analysis

Marquardt–Levenberg algorithm. In the example in Figures 14.2 and 14.3, errors in initial values of the c_j of about 0.2 min. also caused an excessive number of cycles for convergence. Errors larger than this usually caused the convergence to fail. However, with the use of the second derivative to aid in initial guesses of c_j and with careful attention to the choice of initial

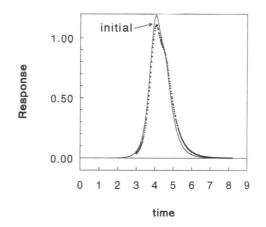

Figure 14.2   Curve from a two-peak model in Table 14.2 (dots) generated with normally distributed absolute noise of $0.5\%A_1$ shown along with the starting line (solid) computed from the initial parameters of a regression analysis (see Table 14.3) onto the same model.

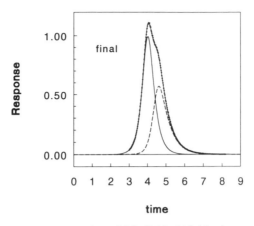

Figure 14.3 Curve from a two-peak model in Table 14.2 (dots) generated with normally distributed absolute noise of 0.5%A_1 shown with computed line (solid) from a successful regression analysis (see Table 14.3 for parameters) onto the same model.

parameters to give a reasonable starting point to the computation (cf., Figure 14.2), the model in Table 14.2 provides excellent convergence.

We end this discussion with a small caveat on the use of the model in Table 14.2. Although this model provides a symmetric peak shape when $a = b$, it is slightly different from Gaussian or Lorentzian shape. This is illustrated by the results of fitting a two-peak model from Table 14.2 onto a set of slightly overlapped peaks. The first peak is Gaussian and the second is constructed from the model itself (Table 14.1). The fit to the first Gaussian

Table 14.3 Initial and Final Parameters for Data in Figures 14.2 and 14.3

Parameter	Initial	True value[a]	Final[b]
A_1	2.2	2.0	1.977 ± 0.014
c_1	4.0	4.0	4.003 ± 0.002
a_1	3.7	4.0	3.980 ± 0.017
b_1	3.1	3.2	3.339 ± 0.068
A_2	0.90	1.0	1.046 ± 0.024
c_2	4.5	4.5	4.495 ± 0.002
a_2	5.1	5.0	4.824 ± 0.111
b_2	2.4	2.0	2.011 ± 0.014
b_5 (fixed)	0	0	0
b_6	0.001	0	$(0.03 \pm 3) \times 10^{-4}$

[a] Data computed with absolute normally distributed error of 0.5%A_1
[b] Value given with standard error from the regression analysis.

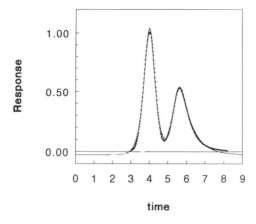

Figure 14.4 Curve from a two-peak model (dots) generated with normally distributed absolute noise of 0.5% of the first peak height. The first peak has a Gaussian shape and the second is computed from the model in Table 14.1. The solid line was computed from the best fit parameters onto the two-peak model in Table 14.2.

peak is not as good as the fit to the second peak computed from the model itself (Figure 14.4). The fitted function is slightly wider at the bottom than the Gaussian peak, and the peak height is slightly overestimated. The baseline offset, which was zero in the computed data set, is underestimated by the regression program. This may be acceptable in some, but probably not all, analyses where Gaussian peak are encountered.

We include the example above to make the point that, although the models in Table 14.1 and 14.2 can provide a variety of shapes including tailing peaks, they may not be appropriate for all chromatographic data. Where they are not, the exponentially modified Gaussian and Gaussian–Lorentzian models should be considered.

A.2. A Sample Application from High-Pressure Liquid Chromatography

In this section, we discuss applications of the model in Table 14.2 to overlapped peaks in high-pressure liquid chromatography. The first example involves the incomplete separation of C60 and C70 buckminsterfullerenes. The two-peak model (Table 14.2, $k = 2$) gave a good fit to the data (Figure 14.5). The peak area ratio of C60/C70 from nonlinear regression was 0.16. This sample was found to be 82.9% C60 and 17.1% C70 for a ratio of 0.21, in reasonable agreement with the above result, by a different HPLC method giving full baseline resolution of the peaks.

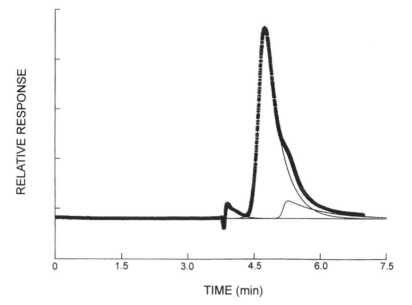

RELATIVE RESPONSE

TIME (min)

Figure 14.5 HPLC chromatogram of C60 and C70 buckminsterfullerenes on a reversed phase C18 column. Mobile phase was hexane at 0.8 mL/min. C60 82.9%; C70 17.1% ratio 4.85. (The authors thank Prof. J. D. Stuart, J. T. Doty, and J. Casavant, University of Connecticut, for providing the original chromatogram.)

In an analytical application of a multipeak model, be it that in Table 14.2, an exponentially modified Gaussian (eq. (14.2)), or any other model, care should be taken to calibrate the analysis properly. We recommend analyzing by nonlinear regression chromatograms of standard mixtures under the same conditions that give the overlapped separation of the sample. Calibration plots of area of component peak vs. concentration of standard can then be constructed for use in determinations with the real samples.

A.3. Overlapped Peaks in Size-Exclusion Chromatography of Complex Polysaccharides

High-performance size exclusion (HPSE) chromatography, also known as *gel permeation chromatography,* is a major tool for the estimation of molecular weights of polymers, proteins, and other macromolecules. Fishman and coworkers [4–6] showed that the HPSE peaks of polysaccharides such as pectins, gums, and starches give good fits to the Gaussian peak shape. Both concentration-sensitive and viscosity detectors are used. The analytical system consists of up to three gel permeation columns in tandem and is subjected to universal calibration with macromolecular standards of

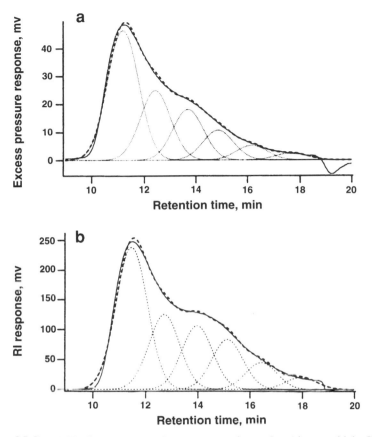

Figure 14.6 HPSE chromatograms for common maize analyzed by a multiple Gaussian model. The solid lines represent experimental data and the dotted lines are computed for the best fit to the model. Response curves for viscosity (top) and refractive index detectors (bottom) are shown. Underlying curves show resolved components. (Reprinted with permission from [6], copyright by Elsevier.)

known molecular weights. These standards were found to give good fits to Gaussian peak shapes. Global values of the intrinsic viscosity, radius of gyration, and weight-average molecular weights of the components can be extracted from the properties of resolved components.

Some examples of this type of analysis for starch from common maize (Figure 14.6) and for gum locust bean (Figure 14.7) show the severely overlapped nature of the HPSE chromatograms typical of complex polysaccharide samples. The overlapped Gaussian model for the analysis of these data is the same as that in Chapter 7. It is clear that without the peak resolution analysis, few results of a quantitative nature would be available for these samples.

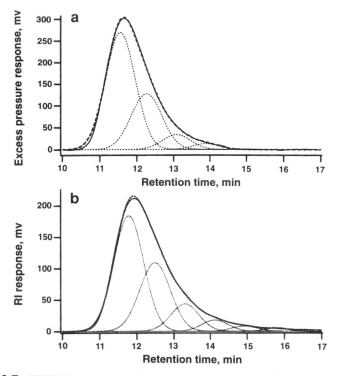

Figure 14.7 HPSE chromatograms for gum locust beans analyzed by a multiple Gaussian model. The solid lines represents experimental data and the dotted lines are computed for the best fit to the model. Response curves for viscosity (a) and refractive index detectors (b) are shown. Underlying curves show resolved components. (Reprinted with permission from [5], copyright by the American Chemical Society.)

B. Multichannel Detection

B.1. Resolution of Time-Resolved Fluorescence Peaks

Conventional spectrophotometric analysis generally measures the signal as dependent variable vs. wavelength or energy as the independent variable. Addition of a second independent variable in the detection system, such as time, for example, greatly increases the information available for each component in a mixture. This is called multichannel or multivariate detection. The larger amount of information present in the data facilitates the mathematical resolution of the signals of the individual components.

An example of a method featuring multivariate detection is time-resolved fluorescence [7]. In the usual algebraic approach used in this book, the model for time resolved fluorescence of a mixture of n components can be expressed as

$$I(t, \lambda) = \sum_{k=1}^{n} A(\lambda)_k C_{0,k} \exp[-t/\tau_k]. \tag{14.3}$$

The symbols in eq. (14.3) are

$I(t,\lambda)$ = measured fluorescence intensity as a function of
λ = wavelength (λ) and time (t),
$A(\lambda)_k$ = proportionality constant between intensity and concentration
   for the kth component in the mixture,
$C_{0,k}$ = concentration of the kth component in the mixture,
τ_k = decay lifetime of the kth component in the mixture.

For a completely unknown system, we would need to estimate the number of components n, and the set of $A(\lambda)_k C_{0,k}$, and τ_k. Knorr and Harris [7] have applied nonlinear regression to the resolution of overlapped peaks in time-resolved fluorescence spectra by using a matrix approach, which is well suited to dealing directly with the multivariate data sets. The approach of these authors is summarized next.

The data consisted of fluorescence spectra of a mixture of n compounds collected at nanosecond time intervals following excitation by a pulse of light. The data were placed in a $w \times t$ matrix $[D]$, with one row for each wavelength observed and the t columns corresponding to the number of time intervals used. A linear response of the detector with changing concentration of each of the n independently emitting compounds was assumed. The elements of this data matrix at wavelength i and time interval j can be written

$$D_{i,j} = \sum_{k=1}^{n} A_{i,k} C_{k,j} \tag{14.4}$$

where $A_{i,k}$ is the molar emission intensity of the kth compound at wavelength i, and $C_{k,j}$ is the concentration of the kth excited state compound at time interval j. The data matrix was expressed as

$$[D] = [A][C] \tag{14.5}$$

where

$[A]$ = $w \times n$ matrix containing the fluorescence spectra of the n compounds,
$[C]$ = $n \times t$ matrix containing the decay of excited state concentrations with time.

Resolution of the spectral data into its individual components requires decomposition of the data matrix into $[A]$ and $[C]$. Assuming first-order decay kinetics for all excited states, a trial matrix $[C']$ was defined by

$$C'_{k,j} = I_j * \exp(-j\Delta t/\tau_k) \tag{14.6}$$

where I_j = measured time response function of the instrument and * represents the convolution of the two functions (see eqs. (14.1) and (14.2)).

Equation (14.6) contains one parameter for each compound, τ_k. Initial guesses of these τ_k provide an intial $[C']$. The best spectral matrix $[A']$ for these parameters is given by

$$[A'] = [D][C']^T\{[C'][C']^T\}^{-1}. \tag{14.7}$$

The $[A']$ found from eq. (14.7) for a given $[C']$ minimized the error sum S expressed in matrix form (Section 2.B.3). This error sum is equivalent to the usual sum of squares of deviations between the experimental data and the model. The model can be expressed as

$$[D'] = [A'][C']. \tag{14.8}$$

The regression analysis employed the n lifetimes as parameters and the minimum in the error sum S was found by using the modified simplex algorithm. Because negative elements in $[A']$ are not physically reasonable, S was multiplied by a function that penalized the occurrence of any such negative elements during the analysis.

The method was evaluated by using noisy simulated data representing a two-component mixture. Real samples containing two fluorescing compounds were also analyzed. Better precision was found for lifetimes and pre-exponential factors when compared to analysis of decay data at a single wavelength [7]. The number of compounds present in the mixture was found by comparing goodness of fit criteria for a series of models with successively increasing integer n values. The n value providing the smallest minimum error sum was taken as the best model (cf. Chapter 6). This type of analysis would also benefit from employing the extra sum of squares F test (Section 3.C.1).

B.2. Resolution of Overlapped Chromatographic Peaks Using Multichannel Detection

Multichannel detectors attached to chromatographic columns are characteristic of the so-called hyphenated analytical techniques, such as gas chromatography–mass spectrometry (GC-MS), gas chromatography–infrared spectrometry (GC-IR), and liquid chromatography–mass spectrometry (GC-MS). A model that has been tested for this purpose employed a Gaussian peak convoluted with an exponential decay (eq. (14.2)) to account for the tailing of chromatographic peaks [7]. The model in Table 14.1 should also be appropriate but has not been tested for this purpose at the time of this writing. The main difference from the single channel problem is that, in addition to time, the multichannel detector data has a second independent variable equivalent to channel number. Therefore, the model in Table 14.2 expressed for multivariate detection would have the following form:

$$R(m, t) = \sum_{j=1}^{k} \left\{ \frac{A_{j,m}}{\exp[-a_j(t - c_j)] + \exp[-b_j(t - c_j)]} \right\} \quad (14.9)$$

where m is the detector channel number, and j denotes the jth component. It is assumed in eq. (14.9) that the peak from each detection channel would have the same shape, as governed by the chromatographic analysis, but that the $A_{j,m}$ will be different because of different sensitivities in the different channels.

Data obtained from multichannel detection of chromatographic peaks has been resolved by matrix regression methods similar to those described for time-resolved fluorescence (cf., Section 14.B.1). The multichannel detector data were organized in a $[D]$ matrix of x rows and t columns, where x is the number of spectral channels and t is the number of chromatographic time channels at which the spectra are observed [8]. The time axis corresponds to the elution time of the chromatographic experiment. For resolution of overlapped chromatographic peaks, the data matrix is decomposed into its component matrices as defined by

$$[D] - [A][Q][C] \quad (14.10)$$

where the component matrices are

$[A]x \times n =$ contains the normalized spectra of the n compounds in the analyte mixture as represented in the overlapped peak,

$[Q]n \times n -$ diagonal matrix representative of the amount of each compound present; and

$[C]n \times t =$ contains normalized chromatograms of each of the n compounds.

The individual components of the overlapped chromatographic peaks were modeled with the convolution of the Gaussian peak shape with a single-sided exponential to account for tailing [8]. Thus, the elements of the trail matrix $[C']$ for the kth compound at time j are given conceptually by

$$C'_{k,j} = G_{k,j} * \exp(-j\Delta t/\tau_k) \quad (14.11)$$

where * represent the convolution of the two functions (cf., eq. (14.2), τ_k is the tailing "lifetime" for the kth compound, and Δt is the time interval between collection of each individual spectra. $G_{k,j}$ is the Gaussian peak shape, expressed in a slightly different way than in Chapters 3 and 7, as

$$G_{k,j} = (t_k/N^{1/2})(2\pi)^{-1/2} \exp\left[\frac{-(j\Delta t - t_k)^2}{2(t_k/N^{1/2})^2} \right]. \quad (14.12)$$

In eq. (14.12), N is the number of theoretical plates in the chromatographic column. The tailing decay lifetimes τ_k were assumed to have been
measured on pure standards. Therefore, only one parameter per compound,
the chromatographic retention time t_k, was required for the nonlinear regression analysis.

Analogous to the use of eq. (14.7), the best matrix product $[A'][Q']$ was
found from

$$[A'][Q'] = [D][C']^T\{[C'][C']^T\}^{-1}. \tag{14.13}$$

The error sum minimized for the regression analysis was the sum of
squares of the deviations between $[D]$ and the model matrix $[D']$ weighted
by the inverse of the degrees of freedom in the analysis [8]. The model is
expressed as

$$[D'] = [A'][Q'][C']. \tag{14.14}$$

As in the time-resolved fluorescence example, the error sum included a
penalty for negative elements in $[A']$. The error sum was minimized using
the simplex algorithm.

The preceding methodology was first evaluated with GC-MS data on
two- and three-component mixtures, all of which eluted in a single, poorly
resolved chromatographic peak [8]. The number of components n was
found accurately by successive fits with increasing integer n values until
the smallest minimum error sum was found. In most cases, the correct n
was found easily by this method.

The regression method successfully resolved major components of test
mixtures having a high degree of overlap in both their chromatograms and
mass spectra. However, quantitative analysis of minor components was
limited in this initial application because of a small spectral sampling rate;
that is, Δt was too large. Nevertheless, the technique avoids some of the
problems associated with arbitrary peak separation methods or those based
on library searching or factor analysis [8].

A second application of the multichannel detection peak resolution
method was made to high-pressure liquid chromotography using a multiple-
wavelength photodiode array as the detector [9]. The data matrix was
expressed by eq. (14.5), where A contained the spectra of the standards.
Chromatographic profiles in the matrix $[C]$ were normalized over the total
elution volume. The model for each component chromatographic peak was
again the Gaussian peak shape convoluted with an exponential tail, as in
eq. (14.11). In this case, parameters used for each k component were retention time (t_k) and peak width $W = t_k/N^{1/2}$. The modified simplex algorithm
was used to find the minimum error sum.

The procedure was tested for resolving five- to eight-component mixtures. Successful resolution of overlapped chromatographic peaks with up
to seven components was achieved. Severely overlapped minor components

in binary mixtures were detected at a 95% level of confidence when present in amounts >35%, depending on peak resolution factors and S/N of the data [9].

References

1. J. M. Gladney, B. F. Dowden, and J. D. Swalen, "Computer Assisted Gas-Liquid Chromatography," *Anal. Chem.* **41** (1969), pp. 883–888.
2. J. P. Foley and J. G. Dorsey, "A Review of the Exponentially-Modified Gaussian Function," *J. Chromatographic Sci.* **22** (1984), pp. 40–46.
3. A. Losev, "On a Model Line Shape for Asymmetric Spectral Peaks," *Appl. Spectrosc.* **48** (1994), pp. 1289–1290.
4. M. L. Fishman, Y. S. El-Atway, S. M. Sodney, D. T. Gillespie, and K. B. Hicks, "Component and Global Average Radii of Gyration of Pectins from Various Sources," *Carbohydrate Polym.* **15** (1991), pp. 89–104.
5. P. T. Hoagland, M. L. Fishman, G. Konja, and E. Clauss, "Size Exclusion Chromatography with Viscocity Detection of Complex Polysaccharides: Component Analysis," *J. Ag. Food Chem.* **41** (1993), pp. 1274–1281 and references therein.
6. M. L. Fishman and P. T. Hoagland, "Characterization of Starches Dissolved in Water by Microwave Heating in a High Pressure Vessel," *Carbohydrate Polym.* **23** (1994), pp. 175–183.
7. F. J. Knorr and J. M. Harris, "Resolution of Multicomponent Fluorescence Spectra by an EmissionWavelength-Decy Time Matrix," *Anal. Chem.* **53** (1981), pp. 272–283.
8. F. J. Knorr, H. R. Thorsheim, and J. M. Harris, "Multichannel Detection and Numerical Resolution of Overlapping Chromatographic Peaks," *Anal. Chem.* **53** (1981), pp. 821–825.
9. S. D. Frans, M. L. McConnell, and J. M. Harris, "Multiwavelength and Reiterative Least Squares Resolution of Overlapped Liquid Chromatographic Peaks," *Anal. Chem.* **57** (1985), pp. 1552–1559.

Appendix I

Linear Regression Analysis

```
                    LINEAR REGRESSION ANALYSIS.
                    ---------------------------
X               Y-OBS          Y-CALC          DIFFERENCE
-               -----          ------          ----------
 1              .011           1.118404E-02                   1.840368E-04
 5              .049           5.096154E-02                   1.96154E-03
10              .102           .1006834        -1.316577E-03
20              .199           .2001272         1.127169E-03
30              .304           .2995709        -4.429072E-03
40              .398           .3990147         1.01468E-03
50              .497           .4984585         1.458436E-03
SLOPE =  9.944376E-03
STANDARD DEVIATION OF SLOPE =  5.379905E-05
INTERCEPT =  1.239661E-03
STANDARD DEVIATION OF INTERCEPT =  1.51158E-03
STANDARD DEVIATION OF REGRESSION =  2.438353E-03
COEFFICIENT OF CORRELATION =  .9999269
10 REM linear regression with sample calculations - linreg2.bas
20 REM ********************** SEPTEMBER 20,1984 ***************************
30 REM ****   N = NUMBER OF DATA POINTS
40 REM ****   J = LOOP COUNTER
50 REM ****   X4 = MEAN VALUE OF X'S
60 REM ****   Y4 = MEAN VALUE OF Y'S
70 REM ****   X1 = TOTAL VALUE OF X'S
80 REM ****   Y1 = TOTAL VALUE OF Y'S
90 REM ****   X2 = TOTAL VALUE OF X^2'S
100 REM ****   X3 = TOTAL CROSS PRODUCT
105 REM INPUT DATA STARTING L 2000 AS # DATA PAIRS, x, y
110 REM ****   S = SLOPE OF THE BEST FITTED LINE
120 REM ****   B = INTERCEPT OF THE LINE
130 PRINT"                   LINEAR REGRESSION"
140 PRINT"*********************************************************************"
150 READ N
155 DIM X(N),Y(N),Y5(N),D(N)
160 FOR J = 1 TO N
170 READ X(J),Y(J)
180 NEXT J
190 GOSUB 210
200 STOP
210 REM *****************************************************************
220 REM *****       THIS SUBROUTINE CALCULATES THE LINEAR REGRESSION   *****
230 REM *****   AND ITS ANALYSIS.                                      *****
240 REM *****************************************************************
250 X1 = 0
260 X2 = 0
270 X3 = 0
280 Q1 = 0
290 Y1 = 0
300 R1 = 0
310 R2 = 0
320 FOR J = 1 TO N
330 X1 = X1 + X(J)
340 Y1 = Y1 + Y(J)
350 X2 = X2 + X(J)^2
360 X3 = X3 + X(J)*Y(J)
370 NEXT J
380 S = ( N * X3 - X1 * Y1 ) / ( N * X2 - X1^2 )
390 B = ( Y1 * X2 - X1 * X3 ) / ( N * X2 - X1^2 )
400 X4 = X1 / N
410 Y4 = Y1 / N
```

```
420 X9 = X2 - N * X4^2
430 FOR J = 1 TO N
440 Y5(J) = S * X(J) + B
450 D(J) = Y5(J) - Y(J)
460 Q1 = Q1 + Y(J)^2
470 NEXT J
480 QX = Q1 - N * Y4^2
490 FOR J = 1 TO N
500 R1 = R1 + ( Y5(J) - Y4 )^2
510 R2 = R2 + ( Y(J) - Y4 )^2
520 NEXT J
530 R = SQR ( R1 /R2 )
540 Q = 0
550 FOR J = 1 TO N
560 Q = Q + D(J)^2
570 S1 = ( SQR ( N / (N * X2 - X1^2 ) ) ) * ( SQR ( Q / ( N - 2 ) ) )
580 S2 = ( SQR ( X2 / ( N * X2 - X1^2 ) ) ) * ( SQR ( Q / ( N - 2 ) ) )
590 NEXT J
600 UR = ( QX - S^2 * X9 ) / (N - 2 )
610 SR = SQR ( UR )
620 PRINT "TYPE 1 FOR LEAST SQUARE FIT ONLY, OR TYPE 0 FOR SAMPLE CALC."
630 INPUT K
640 IF K = 0 THEN 660
650 IF K > 0 THEN 770
660 PRINT "HOW MANY SAMPLES ?"
670 INPUT P
680 PRINT "ENTER AVERAGE RESPONSE AND NUMBER OF MEASUREMENTS FOR EACH"
690 PRINT "INDIVIDUAL SAMPLE."
700 DIM A(P),MN(P),S3(P),SX(P)
710 FOR I = 1 TO P
720 INPUT A(I),MN(I)
730 PRINT
740 S3(I) = ( A(I) - B ) / S
750 SX(I) = SQR ((UR/S^2)*((1/MN(I)+1/N)+(A(I)-Y4)^2/((S^2)*X9)))
760 NEXT I
770 PRINT "                         LINEAR REGRESSION ANALYSIS."
780 PRINT "                         ---------------------------"
790 PRINT "X","Y-OBS","Y-CALC","DIFFERENCE"
800 PRINT "-","-----","------","----------"
810 FOR J = 1 TO N
820 PRINT X(J),Y(J),Y5(J),D(J)
830 NEXT J
840 IF K>= 1 THEN 950
850 PRINT "                      SAMPLE CALCULATION FROM STANDARD CURVE"
860 PRINT "                      -------------------------------------"
870 PRINT "SAMPLE NUMBER","Y-OBS","X-CALC","STD. DEV."
880 PRINT "-------------","-----","------","---------"
890 FOR I = 1 TO P
900 PRINT "   ";I,A(I),S3(I),SX(I)
910 NEXT I
930 PRINT "HIT ANY KEY FOR SUMMARY"
940 V$=INKEY$: IF V$="" THEN 940
950 PRINT " "
960 PRINT "SLOPE = ";S
970 PRINT "STANDARD DEVIATION OF SLOPE = ";S1
980 PRINT "INTERCEPT = ";B
1000 PRINT "STANDARD DEVIATION OF INTERCEPT = ";S2
1010 PRINT "STANDARD DEVIATION OF REGRESSION = ";SR
1020 PRINT "COEFFICIENT OF CORRELATION = ";R
1025 PRINT "FOR HARD COPY, TYPE GOTO 1100"
```

```
1030 RETURN
1100 LPRINT "                        LINEAR REGRESSION ANALYSIS."
1110 LPRINT "                        --------------------------"
1120 LPRINT "X","Y-OBS","Y-CALC","DIFFERENCE"
1130 LPRINT "-","-----","------","----------"
1160 FOR J = 1 TO N
1170 LPRINT X(J),Y(J),Y5(J),D(J)
1180 NEXT J
1190 IF K>= 1 THEN 1300
1200 LPRINT "                    SAMPLE CALCULATION FROM STANDARD CURVE"
1210 LPRINT "                    ------------------------------------"
1220 LPRINT "SAMPLE NUMBER","Y-OBS","X-CALC","STD. DEV."
1230 LPRINT "-------------","-----","------","---------"
1240 FOR I = 1 TO P
1250 LPRINT "   ";I,A(I),S3(I),SX(I)
1260 NEXT I
1290 LPRINT " ":LPRINT "   "
1300 LPRINT "SLOPE = ";S
1310 LPRINT "STANDARD DEVIATION OF SLOPE = ";S1
1320 LPRINT "INTERCEPT = ";B
1340 LPRINT "STANDARD DEVIATION OF INTERCEPT = ";S2
1350 LPRINT "STANDARD DEVIATION OF REGRESSION = ";SR
1360 LPRINT "COEFFICIENT OF CORRELATION = ";R
2000 DATA 7
2010 DATA 1,0.011,5,0.049,10,0.102,20,0.199,30,0.304,40,.398,50,0.497
2020 DATA .8,23.7192,1,28.779
2030 DATA 1.5,40.4789
```

Index